PAULI LECTURES ON PHYSICS VOLUME 2

Optics and the Theory of Electrons

Wolfgang Pauli

Edited by Charles P. Enz

Translated by S. Margulies and H. R. Lewis

Foreword by Victor F. Weisskopf

DOVER PUBLICATIONS, INC.
Mineola, New York

Copyright

Bibliographical Note

This Dover edition, first published in 2000, is an unabridged republication of the work originally published in 1973 by The MIT Press, Cambridge, Massachusetts and London, England.

Library of Congress Cataloging-in-Publication Data

Pauli, Wolfgang, 1900–1958.
[Vorlesung über Optik und Elektronentheorie. English]
Optics and the theory of electrons / Wolfgang Pauli ; edited by Charles P. Enz ; translated by S. Margulies and H.R. Lewis ; foreword by Victor F. Weisskopf.
p. cm. — (Pauli lectures on physics ; v. 2)
Originally published: Cambridge, Mass. : MIT Press, 1973.
Includes bibliographical references and index.
ISBN 0-486-41458-2 (pbk.)
1. Optics. 2. Light, Wave theory of. I. Enz, Charles P. (Charles Paul), 1925– II. Title.

QC3 .P35 2000 vol. 2
[QC355.2]
530 s—dc21
[535]

00-031577

Manufactured in the United States of America
Dover Publications, Inc., 31 East 2nd Street, Mineola, N.Y. 11501

Pauli Lectures on Physics in Dover Editions

Volume 1. Electrodynamics (41457-4)
Volume 2. Optics and the Theory of Electrons (41458-2)
Volume 3. Thermodynamics and the Kinetic Theory of Gases (41461-2)
Volume 4. Statistical Mechanics (41460-4)
Volume 5. Wave Mechanics (41462-0)
Volume 6. Selected Topics in Field Quantization (41459-0)

15. Light entering and leaving crystals 106

5. Molecular Optics 109

16. Dispersion by undamped oscillators 109
17. Dispersion by damped oscillators 117
18. Scattering of light 121
19. Optical activity 133
20. Magneto-optics 143

Supplementary Bibliography 149

Appendix. Comments by the Editor 151

Index 155

Contents

Foreword by Victor F. Weisskopf vii

Preface by the Editor ix

1. Geometrical Optics 1

1. Fermat's principle 1
2. The principles of Malus and Huygens. Laws of image formation 7
3. Hamilton's theory 14
4. Photometry 25

2. Theory of Interference and Diffraction 28

5. General kinematics of waves 28
6. Refraction, reflection, interference 39
7. Theory of diffraction 45

3. Maxwell's Theory 67

8. Foundations of the theory 67
9. Nonabsorbing media (Fresnel's formulae) 71
10. Absorbing media (optics of metals) 77
11. Standing waves 82

4. Crystal Optics 84

12. Relations for the wave normal 84
13. Ray variables 91
14. Singularities 99

15. Light entering and leaving crystals 106

5. Molecular Optics 109

16. Dispersion by undamped oscillators 109
17. Dispersion by damped oscillators 117
18. Scattering of light 121
19. Optical activity 133
20. Magneto-optics 143

Supplementary Bibliography 149

Appendix. Comments by the Editor 151

Index 155

Foreword

It is often said that scientific texts quickly become obsolete. Why are the Pauli lectures brought to the public today, when some of them were given as long as twenty years ago? The reason is simple: Pauli's way of presenting physics is never out of date. His famous article on the foundations of quantum mechanics appeared in 1933 in the German encyclopedia *Handbuch der Physik*. Twenty-five years later it reappeared practically unchanged in a new edition, whereas most other contributions to this encyclopedia had to be completely rewritten. The reason for this remarkable fact lies in Pauli's style, which is commensurate to the greatness of its subject in its clarity and impact. Style in scientific writing is a quality that today is on the point of vanishing. The pressure of fast publication is so great that people rush into print with hurriedly written papers and books that show little concern for careful formulation of ideas. Mathematical and instrumental techniques have become complicated and difficult; today most of the effort of writing and learning is devoted to the acquisition of these techniques instead of insight into important concepts. Essential ideas of physics are often lost in the dense forest of mathematical reasoning. This situation need not be so. Pauli's lectures show how physical ideas can be presented clearly and in good mathematical form, without being hidden in formalistic expertise.

Pauli was not an accomplished lecturer in the technical sense of the word. It was often difficult to follow his courses. But when the sequence of his thoughts and the structure of his logic become apparent, the attentive follower is left with a new and deeper knowledge of essential concepts and with a clearer insight into the splendid architecture of reason, which is theoretical physics. The value of the lecture notes is not diminished by the fact that they were written not by him but by some of his collaborators. They bear the mark of the master in their conceptual structure and their mathematical rigidity. Only here and there does one miss words and comments of the master. Neither does one notice the passing of time in his lectures, with the sole exception of the lectures on field quantization, in which some concepts are formulated in a way that may appear old-fashioned to some today. But even these lectures should be of use to modern students because of their compactness and their direct approach to the central problems.

May this volume serve as an example of how the concepts of theoretical physics were conceived and taught by one of the great men who created them.

Victor F. Weisskopf

Cambridge, Massachusetts

Preface

When Pauli taught optics to the students of ETH at Zürich, the main interest of this subject, as of mechanics, was to prepare the students for the ideas of wave mechanics. In this respect the section in this book on Hamilton's theory and its application to anisotropic media in the section on crystal optics are particularly interesting. But otherwise, optics was at Pauli's time a subject of past glory.

This has radically changed with the advent of the laser, which has again put optics on the frontier of science. Subjects such as nonlinear optics and holography, which this year won a Nobel Prize, have opened up entirely new technologies for which a good knowledge of old-fashioned linear classical optics is a necessary preliminary.

This is a fascinating lesson about the unpredictability of the evolution of science. And it is perhaps the best justification for this English translation. For Pauli was not so much interested in detail as in a critical and logical exposition. This makes the present lectures a concise and rewarding introduction to the subject.

The German notes prepared by A. Scheidegger and published in 1948 were not without flaws. This was the reason that Pauli had encouraged a second version of notes prepared by P. Erdös, now Professor at the University of Florida at Talahassee. If for this translation the first version was chosen after all it was because it reflects better the original spirit and compact style of Pauli. The flaws have been eliminated and some precision has been added in the comments of the appendix. This process of clarification has made the work of the translators worth special mention. As editor I am quite pleased to see that the hard work of all of us has produced an improved version of a course I always liked.

Charles P. Enz

Geneva, 2 November 1971

Optics and the Theory of Electrons

PAULI LECTURES ON PHYSICS VOLUME 2

Chapter 1. Geometrical Optics

1. FERMAT'S PRINCIPLE

If there are two homogeneous, isotropic media, M_1 and M_2, which are characterized by their indices of refraction, n_1 and n_2, then Snell's law holds:

$$n_1 \sin \vartheta_1 = n_2 \sin \vartheta_2 \, . \qquad [1.1]$$

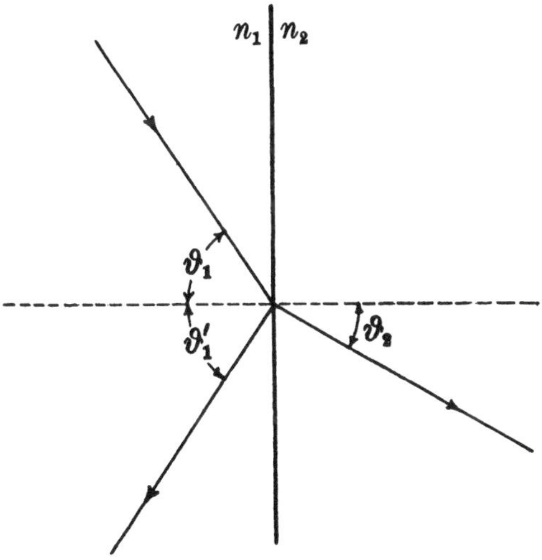

Figure 1.1

For $n_1/n_2 > 1$, as ϑ_1 is increased, a point is reached beyond which there exists no real solution for ϑ_2. At the critical angle defined by

$$\sin \vartheta_1^{(l)} = \frac{n_2}{n_1},$$

the refracted light propagates parallel to the interface between M_1 and M_2. Experiment shows that for $\vartheta_1 > \vartheta^{(l)}$, total reflection occurs. However, some reflection always occurs, whether total or not. In addition to the refracted ray, there is also a reflected ray; that is, the original ray is split into two. For the reflected ray, we have the law

$$\vartheta_1' = \vartheta_1 . \qquad [1.2]$$

The refracted ray as well as the reflected ray lie in the plane of incidence.

That the index of refraction n is independent of the angle of incidence ϑ is characteristic of the isotropy of the medium. On the other hand, it should be well noted that n depends upon the color of the light. In vacuum, by definition, $n = 1$ for all colors. Thus, n is greater than unity in material media.

The law of refraction as well as that of reflection can be characterized by means of an extremal property. If we

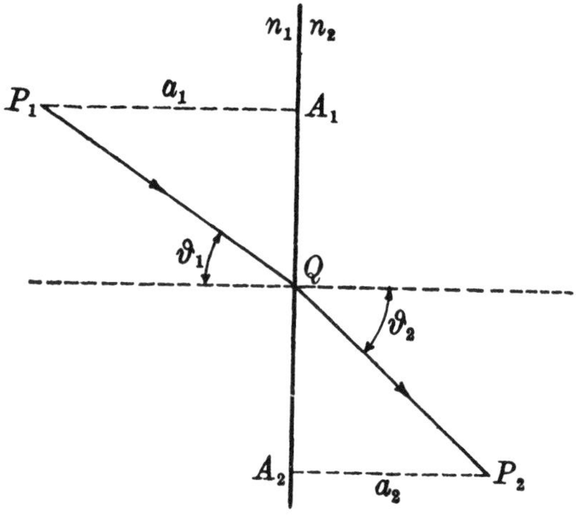

Figure 1.2

denote the product of index of refraction and light path by the optical path L,

$$L = n_1 P_1 Q + n_2 P_2 Q, \tag{1.3}$$

then the principle formulated by Fermat states that a light ray always travels in such a way that

$$\delta L = 0. \tag{1.4}$$

This law is identical with the laws of refraction and reflection given above.

Proof. a. Law of Refraction

From Fig. 1.2 we see that

$$L = n_1 a_1/\cos\vartheta_1 + n_2 a_2/\cos\vartheta_2,$$

$$A_1 A_2 = a_1 \tan\vartheta_1 + a_2 \tan\vartheta_2 = \text{constant (auxiliary condition)}.$$

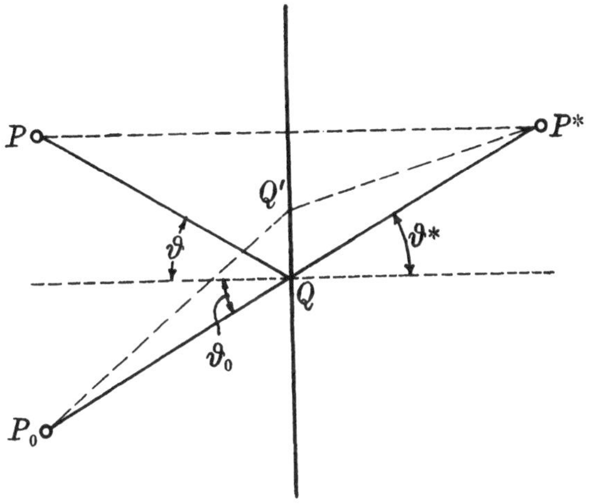

Figure 1.3

Using the method of Lagrangian multipliers, we obtain

$$\delta L + \lambda\,\delta(A_1 A_2) = 0$$

for arbitrary variations of ϑ_1 and ϑ_2. Thus,

$$a_1\left(\frac{n_1 \sin\vartheta_1}{\cos^2\vartheta_1} + \lambda\frac{1}{\cos^2\vartheta_1}\right)\delta\vartheta_1 + a_2\left(\frac{n_2 \sin\vartheta_2}{\cos^2\vartheta_2} + \lambda\frac{1}{\cos^2\vartheta_2}\right)\delta\vartheta_2 = 0,$$

so that

$$n_1 \sin \vartheta_1 + \lambda = 0\,,$$

$$n_2 \sin \vartheta_2 + \lambda = 0\,.$$

Eliminating λ from these equations yields

$$n_1 \sin\vartheta_1 = n_2 \sin\vartheta_2\,.$$

b. Law of Reflection

If P^* is the image of P, then it is seen without computation (see Fig. 1.3) that

$$P_0Q' + Q'P^* > P_0Q + QP^*\,.$$

The optical path is, therefore, minimal.

For curved surfaces, the optical path need not be minimal; it need only be extremal. Later, we shall see that since the velocity of light is proportional to $1/n$, then the optical path is proportional to the transit time of the ray. Thus, according to Fermat, this quantity is to be extremal. However, the concept of transit time becomes meaningful only when considering wave optics.

If we now consider a medium with a continuously variable index of refraction $n = n(x, y, z)$, then we can easily formulate Fermat's principle for this case too; corresponding to Eq. [1.4], we can state the principle as follows: the ray travels between the fixed points P_1 and P_2 in such a manner that

$$\delta S \equiv \delta \int_{P_1}^{P_2} n\, \mathrm{d}s = 0 \qquad (P_1,\ P_2 \text{ fixed})\,. \qquad [1.5]$$

If u is the parameter characterizing a given curve, then

$$S = \int_{u_1}^{u_2} n \left[\sum_i \left(\frac{\mathrm{d}x_i}{\mathrm{d}u}\right)^2\right]^{\frac{1}{2}} \mathrm{d}u\,,$$

where

$$x_i = x_i(u) + \delta x_i(u) \qquad (i = 1, 2, 3)$$

is a neighboring curve, and P_i corresponds to $x(u_i)$. Then,

$$\delta S = \int_{u_1}^{u_2} \sum_k \left\{ \frac{\partial n}{\partial x_k} \delta x_k \left[\sum_i \left(\frac{\mathrm{d}x_i}{\mathrm{d}u} \right)^2 \right]^{\frac{1}{2}} + n \frac{\mathrm{d}x_k}{\mathrm{d}u} \frac{\mathrm{d}}{\mathrm{d}u} (\delta x_k) \cdot \left[\sum_i \left(\frac{\mathrm{d}x_i}{\mathrm{d}u} \right)^2 \right]^{-\frac{1}{2}} \right\} \mathrm{d}u\,.$$

Here, we have used the fact that the differentiations δ and d are independent, so that $\delta\,\mathrm{d}x = \mathrm{d}\,\delta x$. If the second term in the last formula is integrated by parts, then there follows

$$\delta S = \int_{u_1}^{u_2} \sum_k \left\{ \frac{\partial n}{\partial x_k} \left[\sum_i \left(\frac{\mathrm{d}x_i}{\mathrm{d}u} \right)^2 \right]^{\frac{1}{2}} - \frac{\mathrm{d}}{\mathrm{d}u} \left(n \frac{\mathrm{d}x_k}{\mathrm{d}u} \left[\sum_i \left(\frac{\mathrm{d}x_i}{\mathrm{d}u} \right)^2 \right]^{-\frac{1}{2}} \right) \right\} \delta x_k \,\mathrm{d}u$$
$$+ \sum_k n \frac{\mathrm{d}x_k}{\mathrm{d}u} \left[\sum_i \left(\frac{\mathrm{d}x_i}{\mathrm{d}u} \right)^2 \right]^{-\frac{1}{2}} \delta x_k \Bigg|_{P_1}^{P_2}.$$

Since by assumption P_1 and P_2 are fixed, then $\delta x|_{P_1} = \delta x|_{P_2} = 0$, and the last term vanishes. Then, however, δS can vanish for an otherwise arbitrary choice of δx_k only if the integrand in the above formula vanishes identically. If the parameter u is chosen equal to the arc length s along the extremal curve, and if vector notation is used, then this leads to the condition

$$\operatorname{grad} n = \frac{\mathrm{d}}{\mathrm{d}s} \left(n \frac{\mathrm{d}\boldsymbol{x}}{\mathrm{d}s} \right). \qquad [1.6]$$

It might be thought that in the whole derivation the parameter u was superfluous in that it could be set equal to s right from the start. This, however, cannot be done, since while one can indeed choose $u = s$ along the extremal curve, this cannot at the same time be done along the varied comparison curve.

If the variation of S corresponding to a variation of the endpoints P_1 and P_2 is now investigated, subject to the condition that the curve between P_1 and P_2 should remain an extremum, then there follows

$$\delta S = n\left(\frac{\mathrm{d}\boldsymbol{x}}{\mathrm{d}s}\cdot\delta\boldsymbol{x}\right)\Bigg|_{P_1}^{P_2}. \tag{1.7}$$

Here too, Fermat's principle states only that the optical path $\int n\,\mathrm{d}s$ is to be an extremum; what kind of an extremum is not specified. This latter question is of no consequence at all in physics.

If Eq. [1.6] is again decomposed into components, then it should be noted that three independent equations do not result. Since

$$\sum_k\left(\frac{\mathrm{d}x_k}{\mathrm{d}s}\right)^2 = 1\,,$$

a relation must exist between the equations. If one takes the scalar product of Eq. [1.6] with $\mathrm{d}\boldsymbol{x}/\mathrm{d}s$, then there follows

$$\frac{\mathrm{d}\boldsymbol{x}}{\mathrm{d}s}\cdot\operatorname{grad} n = \frac{\mathrm{d}\boldsymbol{x}}{\mathrm{d}s}\cdot\frac{\mathrm{d}}{\mathrm{d}s}\left(n\,\frac{\mathrm{d}\boldsymbol{x}}{\mathrm{d}s}\right) = \frac{\mathrm{d}\boldsymbol{x}}{\mathrm{d}s}\cdot\frac{\mathrm{d}\boldsymbol{x}}{\mathrm{d}s}\,\frac{\mathrm{d}n}{\mathrm{d}s} + \frac{\mathrm{d}\boldsymbol{x}}{\mathrm{d}s}\cdot\frac{\mathrm{d}^2\boldsymbol{x}}{\mathrm{d}s^2}\,n\,,$$

so that $(\mathrm{d}\boldsymbol{x}/\mathrm{d}s)\cdot\operatorname{grad} n = \mathrm{d}n/\mathrm{d}s$, which is clearly trivial geometrically. From this it follows that only the vector components have physical significance. When expanded, Eq. [1.6] yields

$$\nabla n = \frac{\mathrm{d}n}{\mathrm{d}s}\,\frac{\mathrm{d}\boldsymbol{x}}{\mathrm{d}s} + n\,\frac{\mathrm{d}^2\boldsymbol{x}}{\mathrm{d}s^2}\,.$$

One of the well-known Frenet formulas is $\mathrm{d}^2\boldsymbol{x}/\mathrm{d}s^2 = \boldsymbol{e}/R$, where $\boldsymbol{e}$ is a unit vector in the osculating plane perpendicular to $\mathrm{d}\boldsymbol{x}/\mathrm{d}s$, and R is the radius of curvature. It then follows that

$$\left.\begin{aligned}(\operatorname{grad} n)_\perp &= \frac{n}{R}\cdot\boldsymbol{e}\\ \text{or}\qquad \{\operatorname{grad}(\log n)\}_\perp &= \frac{\boldsymbol{e}}{R}\end{aligned}\right\}. \tag{1.8}$$

In the special case where n is a function of only one coordinate, for example $n(x)$, then

$$\frac{\mathrm{d}}{\mathrm{d}s}\left(n\frac{\mathrm{d}y}{\mathrm{d}s}\right)=0\,, \qquad \frac{\mathrm{d}}{\mathrm{d}s}\left(n\frac{\mathrm{d}z}{\mathrm{d}s}\right)=0\,,$$

so that

$$n\frac{\mathrm{d}y}{\mathrm{d}s}=\text{constant}\,, \qquad n\frac{\mathrm{d}z}{\mathrm{d}s}=\text{constant}\,,$$

and the ray remains in the plane $\mathrm{d}y/\mathrm{d}z=\text{constant}$. However, the coordinate system can then be chosen so that the ray lies in the x-y plane. Thus,

$$\mathrm{d}z/\mathrm{d}s=0\,, \qquad n\,\mathrm{d}y/\mathrm{d}s=\text{constant}\,,$$

$$\vartheta=\sphericalangle\left(\frac{\mathrm{d}\boldsymbol{x}}{\mathrm{d}s}\,,x\right), \qquad \frac{\mathrm{d}x}{\mathrm{d}s}=\cos\vartheta\,, \qquad \frac{\mathrm{d}y}{\mathrm{d}s}=\sin\vartheta\,,$$

so that there follows

$$n\sin\vartheta=\text{constant}\,.$$

Thus, Snell's law is again obtained.

From the equations which have been developed here —second-order differential equations with given initial conditions—it is clear that, within the framework of this theory, the splitting of a ray into two can never occur. If a point and an initial direction are specified, then the subsequent course of the ray is completely determined. However, this does not agree with the fact that at the boundary between two media, a splitting of the ray always occurs. From this, it is seen that Fermat's principle can only be a certain approximation to reality.

2. THE PRINCIPLES OF MALUS AND HUYGENS. LAWS OF IMAGE FORMATION

We now wish to investigate the properties of the function S, where

$$S=\int_{P_1}^{P_2} n\,\mathrm{d}s=S(P_1,P_2)\,.$$

From Eq. [1.7] there follows

$$\operatorname{grad} S\cdot\delta\boldsymbol{x}=-n\left(\frac{\mathrm{d}\boldsymbol{x}}{\mathrm{d}s}\cdot\delta\boldsymbol{x}\right),\qquad \text{in } P_1,$$

$$\operatorname{grad} S\cdot\delta\boldsymbol{x}=n\left(\frac{\mathrm{d}\boldsymbol{x}}{\mathrm{d}s}\cdot\delta\boldsymbol{x}\right),\qquad \text{in } P_2,$$

so that

$$\operatorname{grad} S=\mp n\,\frac{\mathrm{d}\boldsymbol{x}}{\mathrm{d}s} \tag{2.1}$$

and

$$|\operatorname{grad} S|^2=n^2\,. \tag{2.2}$$

This last equation is often designated as the eikonal equation, and S as the eikonal. These designations are not always unique in the literature.

Instead of the function S, a related function $\bar{S}$ can be introduced in the following manner: Let F be an arbitrarily specified surface in space,

$$x_k^0=\varphi_k(u,v)\qquad\text{or}\qquad F(\boldsymbol{x}^0)=0\,,$$

and let P be a point external to the surface. Then

$$\bar{S}=\bar{S}(F,P)=\int n\,\mathrm{d}s\,,$$

where the path of integration between F and P is to be taken along that light ray which is perpendicular to F and which passes through P. Our general variational formulae, Eqs. [1.6] and [1.7], then imply that

$$\delta\bar{S}=\int_{P_0}^{P}\left\{\frac{\partial n}{\partial\boldsymbol{x}}-\frac{\mathrm{d}}{\mathrm{d}s}\left(n\,\frac{\mathrm{d}\boldsymbol{x}}{\mathrm{d}s}\right)\right\}\cdot\delta\boldsymbol{x}\,\mathrm{d}s+n\,\frac{\mathrm{d}\boldsymbol{x}}{\mathrm{d}s}\cdot\delta\boldsymbol{x}\Big|_{P_0}^{P}.$$

Here, the first term vanishes because the integration path is taken along an extremum, while the second term vanishes at the lower limit as a consequence of the boundary condition. Thus, for a variation of the point P, there results

$$\operatorname{grad}_P \bar{S} = n \frac{\mathrm{d}\boldsymbol{x}}{\mathrm{d}s} . \qquad [2.3]$$

This is Malus's law. Verbally, it states that if rays have at one time been normal to a surface F, then there always exist surfaces, namely, those for which $\bar{S}$ = constant, which are normal to the rays.

There exists an important geometrical relationship between S and $\bar{S}$ which is known as *Huygens's principle*: *The surface* $\bar{S}$ = constant *is the envelope of the surfaces* $S(P_0, P)$ = constant *associated with each point* P_0 *on* F.

Proof.

Let u, v be the surface parameters. The equation for the envelope of the surfaces S = constant is obtained in the standard way by differentiating with respect to the parameters:

$$\partial S/\partial u = 0, \qquad \partial S/\partial v = 0 ,$$

so that

$$\operatorname{grad}_{P_0} S \cdot \frac{\partial \boldsymbol{x}}{\partial u} = 0 , \qquad \operatorname{grad}_{P_0} S \cdot \frac{\partial \boldsymbol{x}}{\partial v} = 0 ,$$

from which it follows that just those rays which are perpendicular to the given surfaces strike the envelope. Q.E.D.

There are two possibilities for describing phenomena in an isotropic medium. We have the equations

$$\operatorname{grad} n = \frac{\mathrm{d}}{\mathrm{d}s}\left(n \frac{\mathrm{d}\boldsymbol{x}}{\mathrm{d}s}\right)$$

and

$$|\operatorname{grad} S|^2 = n^2 ,$$

whose connection with one another we wish to investigate further. It has already been shown that from solutions of the ordinary differential equation [1.6], solutions of the partial differential equation [2.2] can be constructed. However, the inverse of this is also true: if one has a sufficiently general solution of [2.2], then solutions of [1.6] can be obtained from it.

We consider a two-parameter family of surfaces $F(\boldsymbol{x}^0, a_1, a_2) = 0$ and the corresponding functions

$$\bar{S}(F, P) = \bar{S}(P, a_1, a_2) ,$$

which, because of Eq. [2.3], also satisfy Eq. [2.2]. Upon differentiation with respect to a_1 and a_2, there follows

$$\nabla\bar{S}\cdot\nabla\frac{\partial\bar{S}}{\partial a_1} = 0 , \qquad \nabla\bar{S}\cdot\nabla\frac{\partial\bar{S}}{\partial a_2} = 0 ;$$

that is, the surfaces $\partial\bar{S}/\partial a_{1,2} =$ constant are perpendicular to the surface $\bar{S} =$ constant. If it is assumed that the two-parameter family of surfaces F is so chosen that the surfaces $\partial\bar{S}/\partial a_1 =$ constant and $\partial\bar{S}/\partial a_2 =$ constant do not coincide, then they intersect to form a curve which is perpendicular to $\bar{S} =$ constant and parallel to $\nabla\bar{S}$. This curve is a solution of the ray equation. This may be seen as follows. If we write $\partial\bar{S}/\partial a_i = b_i$, then the equation of such a curve becomes

$$x_k = f_k(s, a_1, a_2, b_1, b_2) .$$

Remembering that $\nabla\bar{S}$ is parallel to $\mathrm{d}\boldsymbol{x}/\mathrm{d}s$ and using Eq. [2.2], we can write $\nabla\bar{S} = n\,\mathrm{d}\boldsymbol{x}/\mathrm{d}s$, from which follows

$$\begin{aligned} n\frac{\mathrm{d}}{\mathrm{d}s}\left(n\frac{\mathrm{d}x_i}{\mathrm{d}s}\right) &= n\sum_k \frac{\mathrm{d}x_k}{\mathrm{d}s}\frac{\partial}{\partial x_k}\frac{\partial\bar{S}}{\partial x_i} = \sum_k \frac{\partial\bar{S}}{\partial x_k}\frac{\partial^2\bar{S}}{\partial x_k\partial x_i} \\ &= \frac{\partial}{\partial x_i}\frac{1}{2}\sum_k\left(\frac{\partial\bar{S}}{\partial x_k}\right)^2 = \frac{\partial}{\partial x_i}\left(\frac{1}{2}n^2\right) = n\frac{\partial n}{\partial x_i} , \end{aligned}$$

so that

$$\frac{\mathrm{d}}{\mathrm{d}s}\left(n\frac{\mathrm{d}\boldsymbol{x}}{\mathrm{d}s}\right) = \operatorname{grad} n ,$$

which is Eq. [1.6]. Q.E.D.

Our calculation shows us a very important relationship between a total differential equation and the partial differential equation associated with it. This connection will be considerably generalized later. The relations were first discovered by Hamilton.

As an application of the variational principles, several theorems regarding the exact imaging of surfaces by means of extended pencils of rays will be discussed here.

a. Imaging of a one-dimensional point manifold by means of a one-dimensional manifold of directions

Object-space quantities will be denoted without primes, image-space quantities with primes. All rays from P are to somehow converge at P' and, similarly, those from Q are to converge at Q'; furthermore, they are to lie in the same plane as $\overrightarrow{PQ} = \delta\boldsymbol{x}$. If the unit tangent vector to the ray is denoted by $\boldsymbol{t}$, then, from Eq. [1.7], the variation laws state that

$$\delta S = S(P, P') - S(Q, Q') = n(\boldsymbol{t} \cdot \delta\boldsymbol{x}) - n'(\boldsymbol{t}' \cdot \delta\boldsymbol{x}') .$$

Thus, with $|\delta\boldsymbol{x}| = \delta l$,

$$\delta S = n\delta l \cos(\boldsymbol{t}, \delta\boldsymbol{x}) - n' \, \delta l' \cos(\boldsymbol{t}', \delta\boldsymbol{x}') .$$

The right side of this equation must be independent of which ray of the pencil is considered. If we denote two rays of the pencil by $\boldsymbol{t}_1$ and $\boldsymbol{t}_2$, then it follows that

$$n\delta l\{\cos(\boldsymbol{t}_1, \delta\boldsymbol{x}) - \cos(\boldsymbol{t}_2, \delta\boldsymbol{x})\} = n'\delta l'\{\cos(\boldsymbol{t}_1', \delta\boldsymbol{x}') - \cos(\boldsymbol{t}_2', \delta\boldsymbol{x}')\} .$$

This can be written vectorially as

$$n(\boldsymbol{t}_1 - \boldsymbol{t}_2) \cdot \delta\boldsymbol{x} = n'(\boldsymbol{t}_1' - \boldsymbol{t}_2') \cdot \delta\boldsymbol{x}' , \qquad [2.4]$$

which is Bruns's law. In accordance with the derivation, the law is, of course, valid only for sufficiently small $\delta\boldsymbol{x}$.

b. Imaging of a two-dimensional point manifold by means of a two-dimensional manifold of directions

The notation is the same as in *a* above. In addition, let E be the plane of $\delta\boldsymbol{x}$ and $\boldsymbol{\tau}$ be the projection of $\boldsymbol{t}$ on E. Then,

$$\boldsymbol{t} \cdot \delta\boldsymbol{x} = \boldsymbol{\tau} \cdot \delta\boldsymbol{x}$$

and, from *a*,

$$n(\boldsymbol{\tau}_1 - \boldsymbol{\tau}_2) \cdot \delta\boldsymbol{x} = n'(\boldsymbol{\tau}_1' - \boldsymbol{\tau}_2') \cdot \delta\boldsymbol{x}' ,$$

$$n(\boldsymbol{\tau}_1 - \boldsymbol{\tau}_3) \cdot \delta_1\boldsymbol{x} = n'(\boldsymbol{\tau}_1' - \boldsymbol{\tau}_3') \cdot \delta_1\boldsymbol{x}' ,$$

where, in addition, the $\boldsymbol{\tau}$'s have been so chosen that $\boldsymbol{\tau}_1 - \boldsymbol{\tau}_2$ and $\boldsymbol{\tau}_1 - \boldsymbol{\tau}_3$ are not parallel. The same is true for $\delta\boldsymbol{x}$ and $\delta_1\boldsymbol{x}$. From the familiar identity

$$(\boldsymbol{A}\times\boldsymbol{B})\cdot(\boldsymbol{C}\times\boldsymbol{D}) = (\boldsymbol{A}\cdot\boldsymbol{C})(\boldsymbol{B}\cdot\boldsymbol{D}) - (\boldsymbol{B}\cdot\boldsymbol{C})(\boldsymbol{A}\cdot\boldsymbol{D}),$$

there follows

$$\begin{aligned}&(\delta\boldsymbol{x}\times\delta_1\boldsymbol{x})\cdot[(\boldsymbol{\tau}_2-\boldsymbol{\tau}_1)\times(\boldsymbol{\tau}_3-\boldsymbol{\tau}_1)]\\ &=[\delta\boldsymbol{x}\cdot(\boldsymbol{\tau}_2-\boldsymbol{\tau}_1)][\delta_1\boldsymbol{x}\cdot(\boldsymbol{\tau}_3-\boldsymbol{\tau}_1)]-[\delta_1\boldsymbol{x}\cdot(\boldsymbol{\tau}_2-\boldsymbol{\tau}_1)][\delta\boldsymbol{x}\cdot(\boldsymbol{\tau}_3-\boldsymbol{\tau}_1)].\end{aligned}$$

Multiplying this by n^2 and using the identities obtained above, we find that the expression

$$n^2([\delta\boldsymbol{x}\times\delta_1\boldsymbol{x}]\cdot[(\boldsymbol{\tau}_2-\boldsymbol{\tau}_1)\times(\boldsymbol{\tau}_3-\boldsymbol{\tau}_1)])$$

is equal to the same expression with primed quantities. The geometric significance of the first vector product appearing in the above expression is that of a surface element defined by $\delta\boldsymbol{x}$ and $\delta_1\boldsymbol{x}$; however, the significance of the second vector product is that of the twice-projected surface area of the solid angle $\mathrm{d}\Omega$ defined by the rays $\boldsymbol{t}_1$, $\boldsymbol{t}_2$, and $\boldsymbol{t}_3$ (for sufficiently small $\Delta\boldsymbol{\tau}$). From this it follows that

$$n^2\,\mathrm{d}f\,\mathrm{d}\Omega\cos(\boldsymbol{N},\boldsymbol{t}) = n'^2\,\mathrm{d}f'\,\mathrm{d}\Omega'\cos(\boldsymbol{N}',\boldsymbol{t}'), \qquad [2.5]$$

where $\boldsymbol{N}$ is the normal to the surface $\mathrm{d}f$. This equation is known as the Kirchhoff-Clausius law.

c. Application of the two theorems to rotationally symmetric systems

1. Let surfaces perpendicular to the symmetry axis be imaged in the form of similar surfaces by means of a pencil of rays which has a finite opening angle $2u$ and which contains the axial ray.

Consider a section through the axis. Writing y instead of $\delta\boldsymbol{x}$ and applying Eq. [2.4] with $\boldsymbol{t}_1$ and $\boldsymbol{t}_2$ as the extreme rays of the pencil, we obtain the important *sine condition*:

$$ny\sin u = n'y'\sin u'. \qquad [2.6]$$

This condition is always satisfied whenever the geometrical assumption which was made applies. The sine condition can also be derived from Kirchhoff's law of Eq. [2.5].

2. Let sections of the axis be imaged by a pencil of opening angle $2u$. If we take $\boldsymbol{t}_1$ as the outermost ray of the pencil and $\boldsymbol{t}_2$ as the axial ray, then it follows from Eq. [2.4] that

$$nl(1-\cos u)=n'l'(1-\cos u')\,. \qquad [2.7]$$

In general, conditions [2.6] and [2.7] cannot both be satisfied. This would be possible only for $u=u'$, $ny=n'y'$ and $nl=n'l'$. In practice (for lenses) this is not the case. However, both conditions can be approximately satisfied when u and u' are small.

d. Imaging of a surface by means of rays within it, where the lines connecting exactly imaged points are the rays themselves

Let P_1, P_2, and P_3 be exactly imaged points. From

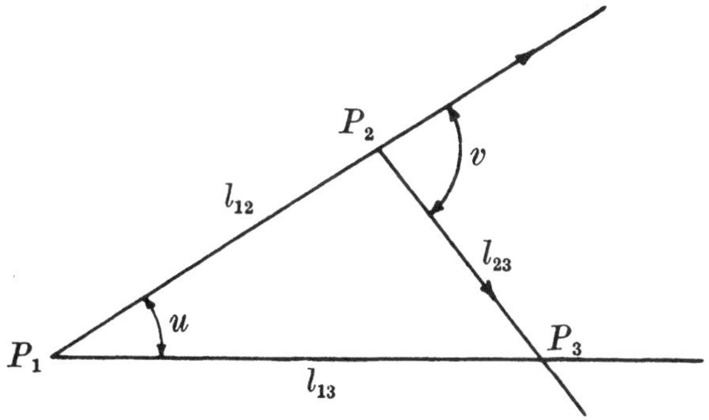

Figure 2.1

Eq. [2.7], it follows that

$$nl_{12}(1-\cos u)=n'l'_{12}(1-\cos u')\,.$$

On the other hand, we also have

$$nl_{13}(1-\cos u)=n'l'_{13}(1-\cos u')$$

and

$$nl_{23}(1-\cos v) = n'l'_{23}(1-\cos v') ,$$

so that

$$l_{12}:l_{13}:l_{23} = l'_{12}:l'_{13}:l'_{23} .$$

Thus, the triangles are similar and their corresponding angles are, therefore, equal:

$$u = u' , \qquad v = v' , \qquad nl = n'l' .$$

3. HAMILTON'S THEORY

a. Special homogeneous form (isotropic bodies)

In geometrical optics, we have become acquainted with the fundamental extremal principle of Fermat

$$\delta \int_{P_1}^{P_2} n \, \mathrm{d}s = 0 .$$

On the other hand, we know that in the mechanics of a mass point the trajectory is determined by the extremal principle

$$\delta \int_{P_1}^{P_2} \sqrt{2m(E - V(x))} \, \mathrm{d}s = 0 . \qquad [3.1]$$

Consequently, if we set $n^2 = 2m(E-V)$, then we obtain a transformation scheme between the equations of mechanics and those of optics such that with each optical problem there is associated a mechanical problem. However, the analogy does not refer to the time development of the phenomena, but, rather, it connects the *trajectories* of a mass point with given initial energy in a potential field and those of a *light ray* in a medium with a given index of refraction (as a function of position).

Thus, in optics as well as in mechanics, the extremal

principle of Eq. [1.5] can be expressed in canonical form. If we set

$$\boldsymbol{p} = n\frac{\mathrm{d}\boldsymbol{x}}{\mathrm{d}s}, \tag{3.2}$$

then

$$\boldsymbol{p}\cdot\frac{\mathrm{d}\boldsymbol{x}}{\mathrm{d}s} = n .$$

With v as a parameter, there then follows

$$0 = \delta S = \delta\int n\,\mathrm{d}s = \delta\int \boldsymbol{p}\cdot\mathrm{d}\boldsymbol{x} = \delta\int \boldsymbol{p}\cdot\frac{\mathrm{d}\boldsymbol{x}}{\mathrm{d}v}\,\mathrm{d}v$$

$$= \int\left\{\delta\boldsymbol{p}\cdot\frac{\mathrm{d}\boldsymbol{x}}{\mathrm{d}v} + \boldsymbol{p}\cdot\frac{\mathrm{d}}{\mathrm{d}v}\,\delta\boldsymbol{x}\right\}\mathrm{d}v = \int\left(\delta\boldsymbol{p}\cdot\frac{\mathrm{d}\boldsymbol{x}}{\mathrm{d}v} - \frac{\mathrm{d}\boldsymbol{p}}{\mathrm{d}v}\cdot\delta\boldsymbol{x}\right)\mathrm{d}v + \boldsymbol{p}\cdot\delta\boldsymbol{x}\Big|_{P_1}^{P_2} .$$

Now, however, it must be noted that $\delta\boldsymbol{x}$ and $\delta\boldsymbol{p}$ can no longer vary independently. Because of Eq. [3.2], we must have

$$\delta G = 0, \qquad \text{where} \qquad G = \tfrac{1}{2}(\boldsymbol{p}^2 - n^2) . \tag{3.3}$$

Then, using the method of Lagrangian multipliers, it follows from $\delta S = 0$ that

$$\frac{\mathrm{d}\boldsymbol{x}}{\mathrm{d}v} = \lambda\frac{\partial G}{\partial \boldsymbol{p}}, \qquad \frac{\mathrm{d}\boldsymbol{p}}{\mathrm{d}v} = -\lambda\frac{\partial G}{\partial \boldsymbol{x}}$$

or, after substitution,

$$\frac{\mathrm{d}\boldsymbol{x}}{\mathrm{d}v} = \lambda\boldsymbol{p}, \qquad \frac{\mathrm{d}\boldsymbol{p}}{\mathrm{d}v} = \lambda n\frac{\partial n}{\partial \boldsymbol{x}} . \tag{3.4}$$

Obviously, these are the same equations as [3.2]. By comparison there follows

$$\left(\frac{\mathrm{d}\boldsymbol{x}}{\mathrm{d}v}\right)^2 = \lambda^2 n^2, \qquad \text{from which} \qquad \lambda n = \frac{\mathrm{d}s}{\mathrm{d}v} .$$

Substituted into Eq. [3.4], this yields the canonical equations

$$n\frac{\mathrm{d}\boldsymbol{x}}{\mathrm{d}s} = \boldsymbol{p}, \qquad \frac{\mathrm{d}\boldsymbol{p}}{\mathrm{d}s} = \frac{\partial n}{\partial \boldsymbol{x}} . \tag{3.5}$$

If $\boldsymbol{p}$ is eliminated between these two equations, then the familiar ray equation [1.6] results. On the basis of our analogy, everything proceeds exactly as in mechanics. If we set $n^2 = 2m(E-V)$, then $\boldsymbol{p}$ becomes the mechanical momentum and $\delta G = 0$ becomes the law of conservation of energy, as can easily be verified by substitution.

Furthermore, we note that mathematically the three-dimensionality is of no consequence in the calculations. In fact, we know that the whole formalism can be carried out in exactly the same way in the mechanics of N mass points (in which there are $3N$ dimensions). In optics, however, only three dimensions occur.

b. Most general homogeneous form (crystal optics)

Furthermore, it is also seen that the specific form of the function n in $\delta\int n\,\mathrm{d}s = 0$ is unessential. Thus, in complete generality we will set

$$\delta S = \delta \int_{P_1}^{P_2} L\,\mathrm{d}u = 0\,, \qquad [3.6]$$

where u is an arbitrary parameter, and L is a function which satisfies the following conditions:

$$L = L\Bigl(q_1, \ldots, q_{f+1}, \underbrace{\frac{\mathrm{d}q_1}{\mathrm{d}u}}_{q_1'}, \ldots, \underbrace{\frac{\mathrm{d}q_{f+1}}{\mathrm{d}u}}_{q_{f+1}'}\Bigr)\,,$$

and

$$L(q_1, \ldots, q_{f+1}, \lambda q_1', \ldots, \lambda q_{f+1}') = \lambda L(q_i, q_i')\,. \qquad [3.7]$$

The requirement of Eq. [3.7] is necessary because the parameter u is completely arbitrary, and it must be demanded that the variational principle of Eq. [3.6] be invariant with respect to changing the parameter. In the case of optics we have $f = 2$. Euler's homogeneity relation corresponding to Eq. [3.7] is

$$\sum q_k' \frac{\partial L}{\partial q_k'} = L\,. \qquad [3.8]$$

Again, in analogy to Eq. [3.2], we set

$$p_k = \frac{\partial L}{\partial q_k'} . \qquad [3.9]$$

From Eqs. [3.8] and [3.9] there then follows

$$L = \sum p_k q_k' , \qquad 0 = \delta S = \delta \int \left(\sum p_k q_k'\right) du . \qquad [3.10]$$

Because L is homogeneous of the first degree in the q_k' (according to Eq. [3.7]), then the p_k are homogeneous in the zeroth degree. If the first f equations of [3.9] are solved for q_k' $(k = 1, 2, 3, \ldots, f)$ and these then substituted into the $(f+1)$th equation of [3.9], then there follows a relation of the form

$$p_{f+1} = p_{f+1}(q_1, \ldots, q_{f+1}, p_1, \ldots, p_f, q_{f+1}') .$$

Because of the homogeneity (of zeroth order) of p_{f+1} in q_{f+1}', it follows that p_{f+1}' cannot depend upon q_{f+1}'. Thus, analogous to Eq. [3.3], there must exist a relationship

$$G(p_1, \ldots, p_{f+1}, q_1, \ldots, q_{f+1}) = 0 . \qquad [3.11]$$

The variational principle [3.10], together with the auxiliary condition [3.11], determines the course of the ray. The variation becomes

$$\delta \int L \, du = \delta \int \sum_k p_k q_k' \, du = \sum_k \int \left[\delta p_k \frac{dq_k}{du} + p_k \frac{d}{du} \delta q_k \right] du$$

$$= \sum_k \int \left(\delta p_k \frac{dq_k}{du} - \frac{dp_k}{du} \delta q_k \right) du + p_k \delta q_k \Big|_{P_1}^{P_2} . \qquad [3.12]$$

With the auxiliary condition [3.11] there finally follows

$$q_k' = \lambda \frac{\partial G}{\partial p_k} , \qquad p_k' = -\lambda \frac{\partial G}{\partial q_k} . \qquad [3.13]$$

On the other hand, it follows quite generally from Eq. [3.6] that

$$\frac{\mathrm{d}}{\mathrm{d}u}\frac{\partial L}{\partial q_k'} = \frac{\partial L}{\partial q_k}.$$

(Here, there is no auxiliary condition because we have worked with the original number of variables without introducing the p_k's.) Thus, the last equation of [3.13] can also be written as

$$p_k' = \frac{\partial L}{\partial q_k} = -\lambda \frac{\partial G}{\partial q_k}. \qquad [3.14]$$

Equations [3.13] and [3.14] are called the canonical equations (analogous to Eqs. [3.4] and [3.5]).

Here too we can consider the surfaces S=constant (wave surfaces), where

$$S = \int_{P_1}^{P_2} L\,\mathrm{d}u.$$

If, in the variational principle of Eq. [3.6], P_1 is held fixed while P_2 is allowed to vary on the surface S= constant, then we must have $\delta S = 0$ by construction (the integral is taken along an extremum). However, from the general variational formula [3.12], it then follows that

$$\sum_k \frac{\partial L}{\partial q_k'}\delta q_k = 0, \qquad \delta q \text{ on } S = \text{constant}. \qquad [3.15]$$

This is the so-called *transversality condition.* In general, this will not be the orthogonality relation [1.7] with $\delta S = 0$, P_1 fixed. Thus, the rays are not perpendicular but only transverse to the surfaces S= constant (see p. 20).

A differential equation can easily be constructed for the function S. From the general variational formula [3.12], it follows that

$$\frac{\partial S}{\partial q_k} = p_k, \qquad [3.16]$$

and this, substituted into Eq. [3.11], yields

$$G\left(\frac{\partial S}{\partial q_1}, \ldots, \frac{\partial S}{\partial q_{f+1}}, q_1, \ldots, q_{f+1}\right) = 0 . \qquad [3.17]$$

If it is now assumed that we have a solution of Eq. [3.17] which depends upon f constants a_ϱ ($\varrho = 1, 2, 3, \ldots, f$), then the theorem that the equations

$$\frac{\partial S}{\partial a_\varrho} = b_\varrho \qquad [3.18]$$

determine light rays is valid. This is the generalization of the relation between a total differential equation and its associated partial differential equation which was mentioned earlier.

Proof.

From Eq. [3.18],

$$q_k = \varphi_k(u, a_1, \ldots, a_f, b_1, \ldots, b_f) \qquad [3.19]$$

(where we have tacitly assumed that the system of equations [3.18] is solvable for the q_k; that is, the matrix $\partial^2 S/\partial a_\varrho \partial q_k$ is of rank f). If this is substituted into Eq. [3.16], there follows

$$\frac{\partial S}{\partial q_k} = p_k = \psi_k(a_1, \ldots, a_f, b_1, \ldots, b_f, u) . \qquad [3.20]$$

Equation [3.17], considered as a function of the q's and a's, when differentiated with respect to a_ϱ yields

$$\sum_k \frac{\partial G}{\partial p_k} \frac{\partial^2 S}{\partial q_k \partial a_\varrho} = 0 . \qquad [3.21]$$

On the other hand, it follows from Eq. [3.18] differentiated with respect to u that

$$\sum_k \frac{\partial^2 S}{\partial a_\varrho \partial q_k} \frac{dq_k}{du} = 0 . \qquad [3.22]$$

These can be simultaneously satisfied only if

$$\frac{\mathrm{d}q_k}{\mathrm{d}u} = \lambda \frac{\partial G}{\partial p_k}. \qquad [3.13a]$$

Similarly, it follows from Eq. [3.17] when differentiated with respect to q that

$$\left.\frac{\partial G}{\partial q_k}\right|_{p\ \text{fixed}} + \sum_l \frac{\partial G}{\partial p_l} \frac{\partial^2 S}{\partial q_l \partial q_k} = 0\,.$$

Substituting into $\mathrm{d}p/\mathrm{d}u$ (from Eq. [3.20]), we obtain

$$\frac{\mathrm{d}p_k}{\mathrm{d}u} = \sum_l \frac{\partial^2 S}{\partial q_k \partial q_l} \frac{\mathrm{d}q_l}{\mathrm{d}u} = \lambda \sum_l \frac{\partial^2 S}{\partial q_k \partial q_l} \frac{\partial G}{\partial p_l} = -\lambda \frac{\partial G}{\partial q_k}. \qquad [3.13b]$$

However, the solutions of Eq. [3.17] given by Eqs. [3.18], [3.19], and [3.20] satisfy the canonical equations [3.13], which, as shown previously, are equivalent to the ray equation. Q.E.D.

Instead of S, we can here again consider a related function defined as follows:

$$\bar{S}(P) = \int_{P_0}^{P} L\,\mathrm{d}u\,,$$

where P is given and P_0 is to lie on a hypersurface $F(q) =$ constant. The path of integration is to be taken along that ray which passes through P and which intersects the surface $F =$ constant transversally (condition [3.15]). Huygens's principle again follows, in which $\bar{S} =$ constant is the envelope of the surfaces $S =$ constant corresponding to the points of the given surface. (The proof is analogous to that used for Huygens's principle in Section 2.)

The theory given in this subsection is to be applied to light propagation in crystals (see Section 13). It should be noted that in this case, "wave surfaces" ($S =$ constant) can still be defined. Although the "wave normal" $\boldsymbol{p} = \{p_1, \ldots\}$ is indeed normal to this surface, it is in general not parallel to the ray direction $\boldsymbol{q}' = \{q_1', \ldots\}$.

c. Inhomogeneous form of Hamilton's theory

In the variational principle of Eq. [3.6], the coordinate q_{f+1} is introduced as the parameter u. Then,

$$q'_k = \frac{\mathrm{d}q_k}{\mathrm{d}u}, \qquad \dot{q}_\alpha = \frac{\mathrm{d}q_\alpha}{\mathrm{d}q_{f+1}} = \frac{q'_\alpha}{q'_{f+1}}. \tag{3.23}$$

The indices have the following ranges: $\alpha = 1, 2, 3, \ldots, f$; $i, k = 1, 2, 3, \ldots, f+1$. Since $L(q', q)$ is homogeneous of the first degree in q', there follows

$$L = \bar{L}(q_1, \ldots, q_{f+1}, \dot{q}_1, \ldots, \dot{q}_f)\, q'_{f+1}, \tag{3.24}$$

$$L\mathrm{d}u = \bar{L}\mathrm{d}q_{f+1}. \tag{3.25}$$

Then,

$$p_\alpha = \frac{\partial L}{\partial q'_\alpha} = \frac{\partial \bar{L}}{\partial \dot{q}_\alpha},$$

and there follows (from [3.23] and [3.24])

$$p_{f+1} = \frac{\partial L}{\partial q'_{f+1}} = \bar{L} - \sum_\alpha \dot{q}_\alpha \frac{\partial \bar{L}}{\partial \dot{q}_\alpha} \equiv -H,$$

$$H = \sum_\alpha \dot{q}_\alpha p_\alpha - \bar{L}, \qquad p_{f+1} + H = 0. \tag{3.26}$$

The canonical equations [3.13] can be rewritten by introducing the function H. The second equation of [3.26] plays the role of the auxiliary condition [3.11], so that it becomes possible to eliminate the parameter λ from Eq. [3.13]:

$$G \equiv p_{f+1} + H(p_\alpha, q_\alpha; q_{f+1}) = 0. \tag{3.27}$$

From Eq. [3.13]

$$\frac{\mathrm{d}q_\alpha}{\mathrm{d}u} = \lambda \frac{\partial H}{\partial p_\alpha}, \qquad \frac{\mathrm{d}p_\alpha}{\mathrm{d}u} = -\lambda \frac{\partial H}{\partial q_\alpha},$$

$$\frac{\mathrm{d}q_{f+1}}{\mathrm{d}u} = \lambda, \qquad \frac{\mathrm{d}p_{f+1}}{\mathrm{d}u} = -\lambda \frac{\partial H}{\partial q_{f+1}}.$$

Eliminating λ, we get the inhomogeneous form of the

canonical equations:

$$\dot{q}_\alpha = \frac{\partial H}{\partial p_\alpha}, \qquad \dot{p}_\alpha = -\frac{\partial H}{\partial q_\alpha}. \tag{3.28}$$

Furthermore, it follows that

$$\frac{\partial H}{\partial q_{f+1}} = \frac{\mathrm{d}H}{\mathrm{d}q_{f+1}}. \tag{3.29}$$

The connection between the total differential equation, the ray equation

$$\frac{\mathrm{d}}{\mathrm{d}q_{f+1}}\left(\frac{\partial \bar{L}}{\partial \dot{q}_\alpha}\right) = \frac{\partial \bar{L}}{\partial q_\alpha},$$

and its associated partial differential equation, is also obtained in the inhomogeneous case, and indeed by again specializing the auxiliary condition [3.11] to that of [3.27]. Substituting $p_k = \partial S/\partial q_k$ into Eq. [3.27], there follows

$$\frac{\partial S}{\partial q_{f+1}} + H\left(\frac{\partial S}{\partial q_\alpha}, q_\alpha, q_{f+1}\right) = 0.$$

Thus, $S = S(q_i, a_\alpha)$ can be determined, and

$$\frac{\partial S}{\partial a_\alpha} = b_\alpha$$

are again the equations for the rays. Of course, upon substitution in this manner, H will also be dependent upon the f constants a_α.

Isotropic optical systems. The inhomogeneous form is practical for application to these systems ($f = 2$):

$$\bar{L} = n(1 + \dot{x}_1^2 + \dot{x}_2^2)^{\frac{1}{2}}, \text{ or } \qquad \bar{L}\,\mathrm{d}x_3 = n\,\mathrm{d}s,$$

$$p_\alpha = \frac{\partial \bar{L}}{\partial \dot{x}_\alpha} = n\dot{x}_\alpha(1 + \dot{x}_1^2 + \dot{x}_2^2)^{-\frac{1}{2}} = n\,\mathrm{d}x_\alpha/\mathrm{d}s;$$

hence

$$n^2 - p_1^2 - p_2^2 = n^2/(1 + \dot{x}_1^2 + \dot{x}_2^2)$$

(which can be verified by substitution), and

$$\bar{L} = n^2(n^2 - p_1^2 - p_2^2)^{-\frac{1}{2}},$$

$$H = -\sqrt{n^2 - p_1^2 - p_2^2} = -p_3.$$

We introduce the characteristic function S,

$$\frac{\partial S}{\partial x_k} = p_k = n\frac{\mathrm{d}x_k}{\mathrm{d}s} \qquad (k = 1, 2, 3)\,,$$

and perform the transformation

$$V(\boldsymbol{p}^0, \boldsymbol{p}) = S - \boldsymbol{p}\;(\boldsymbol{x} - \boldsymbol{a}) + \boldsymbol{p}^0\cdot(\boldsymbol{x}^0 - \boldsymbol{a}^0)\,,$$

where $\boldsymbol{a}$ and $\boldsymbol{a}^0$ are arbitrary [A-1][1] and

$$\mathrm{d}S = \boldsymbol{p}\cdot\mathrm{d}\boldsymbol{x} - \boldsymbol{p}^0\cdot\mathrm{d}\boldsymbol{x}^0\,.$$

Then,

$$\mathrm{d}V = \boldsymbol{p}\cdot\mathrm{d}\boldsymbol{x} - \boldsymbol{p}^0\cdot\mathrm{d}\boldsymbol{x}^0 - \boldsymbol{p}\cdot\mathrm{d}\boldsymbol{x} + \boldsymbol{p}^0\cdot\mathrm{d}\boldsymbol{x}^0 - (\boldsymbol{x} - \boldsymbol{a})\cdot\mathrm{d}\boldsymbol{p} + (\boldsymbol{x}^0 - \boldsymbol{a}^0)\cdot\mathrm{d}\boldsymbol{p}^0 = -(\boldsymbol{x} - \boldsymbol{a})\cdot\mathrm{d}\boldsymbol{p} + (\boldsymbol{x}^0 - \boldsymbol{a}^0)\cdot\mathrm{d}\boldsymbol{p}^0$$

and

$$\frac{\partial V}{\partial \boldsymbol{p}} = -(\boldsymbol{x} - \boldsymbol{a})\,, \qquad \frac{\partial V}{\partial \boldsymbol{p}^0} = \boldsymbol{x}^0 - \boldsymbol{a}^0\,.$$

We go back to the inhomogeneous form

$$p_3 = \sqrt{n^2 - p_1^2 - p_2^2} = -H\,.$$

Writing

$$W(p_1^0, p_2^0, p_1, p_2) \equiv V\left(p_1^0, p_2^0, \sqrt{n^{0 2} - p_1^{0 2} - p_2^{0 2}}, p_1, p_2, \sqrt{n^2 - p_1^2 - p_2^2}\right),$$

we have

$$\mathrm{d}W = -\sum_\alpha X_\alpha \mathrm{d}p_\alpha + \sum_\alpha X_\alpha^0 \mathrm{d}p_\alpha^0\,,$$

with

$$X_\alpha = x_\alpha - a_\alpha + \frac{\partial p_3}{\partial p_\alpha}(x_3 - a_3) = x_\alpha - a_\alpha - \frac{p_\alpha}{\sqrt{n^2 - p_1^2 - p_2^2}}(x_3 - a_3),$$

and the analogous relation for X_α^0.

Axially symmetric systems (*rotational symmetry*). Let the z axis be the axis of rotation. We set

$$\left.\begin{array}{ll} x_1 = x\,, & \mathrm{d}x/\mathrm{d}s = u \\ x_2 = y\,, & \mathrm{d}y/\mathrm{d}s = v \\ x_3 = z\,, & \mathrm{d}z/\mathrm{d}s = w \end{array}\right\} \quad \text{(direction cosines)}\,.$$

[1] Comments [A-1]–[A-7] appear in the Appendix on pp. 151–152.

The arbitrary points $\boldsymbol{a}^0$ and $\boldsymbol{a}$ are chosen along the z axis.

According to our general formulas for isotropic bodies,

$$\frac{\partial W}{\partial u} = -nX, \qquad \frac{\partial W}{\partial v} = -nY,$$

$$\frac{\partial W}{\partial u^0} = n^0 X^0, \qquad \frac{\partial W}{\partial v^0} = n^0 Y^0,$$

with

$$X = x - \frac{nu}{n\sqrt{1-u^2-v^2}}(z-a) = x - \frac{\mathrm{d}x}{\mathrm{d}z}(z-a)\,.$$

This shows that (X, Y) and (X^0, Y^0) represent the coordinates of the points of intersection of the ray with planes perpendicular to the axis at a and a^0, respectively.

The geometric significance of the functions W and V follows directly from the analytic definition

$$W = S - n\boldsymbol{t}\cdot(\boldsymbol{x}-\boldsymbol{a}) + n^0\boldsymbol{t}^0\cdot(\boldsymbol{x}^0-\boldsymbol{a}^0)$$

(where $\boldsymbol{t}$ is a unit vector parallel to the ray). In Fig. 3.1,

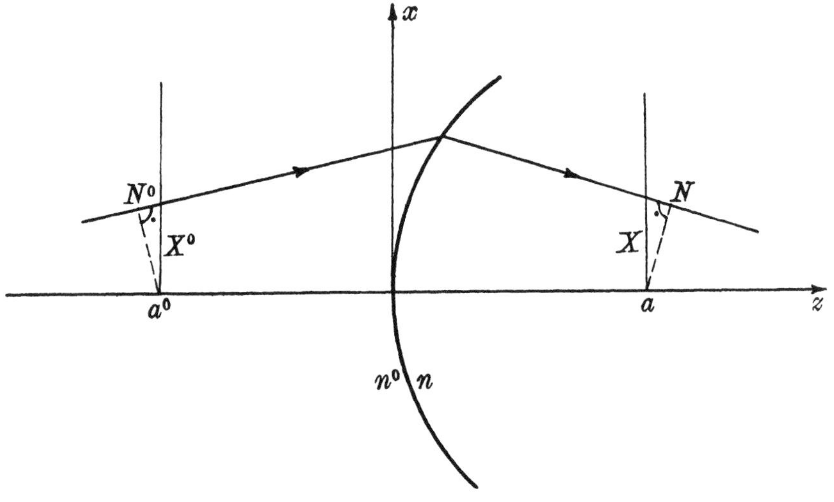

Figure 3.1

this is clearly the optical path from N^0 to N. The functions $V(\boldsymbol{p}^0, \boldsymbol{p})$ and $W(p_1^0, p_2^0, p_1, p_2)$ are called *angular eikonals.*

4. PHOTOMETRY

We start from the assumption that for every optical process (with the exception of emission and absorption) the total amount of light L is conserved. This quantity is also conserved when areas which are bounded by light rays are considered. From these premises, Lambert and Kepler arrived at the following laws.

Let F be the radiating surface and f the irradiated surface. Then $\mathrm{d}L = K\,\mathrm{d}\Omega$, where K is a constant dependent

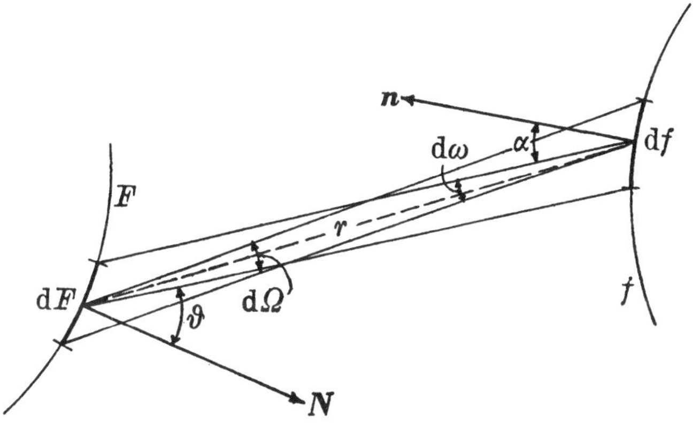

Figure 4.1

upon F. If $\mathrm{d}f$ is an element of the irradiated surface, then

$$\mathrm{d}\Omega = \frac{\mathrm{d}f \cos\alpha}{r^2},$$

$$E = \frac{K\cos\alpha}{r^2}, \qquad \mathrm{d}L = E\,\mathrm{d}f. \tag{4.1}$$

Thus, two surface elements appear equally bright if E is

the same for both; E is called the *intensity of illumination* (or *illuminance*). To be sure, difficulties are encountered in practice when attempting to establish equal brightness for different colors.[2]

Lambert proposed an additional, although problematical, law: the dependence of K on the direction of the light emission is

$$K = I\,\mathrm{d}F\cos\vartheta\,, \qquad [4.2]$$

where I is called the *specific light intensity* (or *photometric brightness*). Indeed, it can be said that Eq. [4.2] is purely formal, since $I = I(\vartheta)$. Only for the case of radiation at thermal equilibrium is $\mathrm{d}I/\mathrm{d}\vartheta = 0$ (isotropic radiator). This case corresponds to the fact that a luminous sphere appears to the eye as a luminous disc of uniform brightness.

The combination of both of these laws (Eqs. [4.1] and [4.2]) yields

$$\mathrm{d}L = I\,\mathrm{d}f\,\mathrm{d}F\cos\alpha\cos\vartheta/r^2\,, \qquad [4.3]$$

$$\mathrm{d}\Omega = \mathrm{d}f\cos\alpha/r^2 = \sin\vartheta\,\mathrm{d}\vartheta\,\mathrm{d}\varphi\,.$$

When integrated for constant I, Eq. [4.3] yields the following for the total L radiated into a cone whose opening half-angle is U:

$$L = \int I\,\mathrm{d}f\,\mathrm{d}F\cos\alpha\cos\vartheta/r^2$$

$$= I\,\mathrm{d}F\int_0^{\pi/2}\int_0^{2\pi}\cos\vartheta\sin\vartheta\,\mathrm{d}\vartheta\,\mathrm{d}\varphi = \pi I\,\mathrm{d}F\sin^2 U\,.$$

For $U = \pi/2$,

$$L = \pi I\,\mathrm{d}F\,.$$

Exact optical mapping. From the preceding discussion, the amount of light passing through corresponding surface

[2] See J. H. C. Müller and C. S. M. Pouillet, *Lehrbuch der Physik* (Vieweg Verlag, Braunschweig, 1925–1934), 11th ed., vols. 1–5.

elements must be the same: $L = L'$. Thus,

$$I \, \mathrm{d}f \cos \alpha \, \mathrm{d}\omega = I' \, \mathrm{d}f' \cos \alpha' \, \mathrm{d}\omega' .$$

However, from Eq. [2.5] it follows that

$$n^2 \, \mathrm{d}f \cos \alpha \, \mathrm{d}\omega = n'^2 \, \mathrm{d}f' \cos \alpha' \, \mathrm{d}\omega',$$

so that

$$I/n^2 = I'/n'^2 . \qquad [4.4]$$

Chapter 2.
Theory of Interference and Diffraction

5. GENERAL KINEMATICS OF WAVES

There exist optical phenomena which contradict the geometrical ray theory presented so far, and which, in addition, cannot be explained by means of a corpuscular theory of light. Such phenomena always appear where the index of refraction varies very rapidly (but perhaps continuously). Among these is the previously considered case of the splitting of a ray into refracted and reflected rays, a phenomenon which cannot be explained, however, by mathematical epsilontics of surface discontinuities. In addition, the diffraction and interference effects familiar from experimental physics belong to this category of phenomena.

All of these phenomena can be treated with the help of a wave concept in which the transversality that is specific to light will be ignored for the time being. The quantitative formulation of this theory originated with Young, Fresnel, and Euler.

a. Oscillations

The *simplest form* of a harmonic oscillation is

$$x = A \cos(\omega t + \delta) ,$$

where $\omega = 2\pi\nu$ is the circular frequency, and ν is the number of oscillations per second. The principle of superposition holds:

$$A_1 \cos(\omega t + \delta_1) + A_2 \cos(\omega t + \delta_2) = A \cos(\omega t + \delta),$$

with the relations

$$A_1 \cos\delta_1 + A_2 \cos\delta_2 = A \cos\delta,$$

$$A_1 \sin\delta_1 + A_2 \sin\delta_2 = A \sin\delta,$$

from which A and δ can be determined. The formulas are simpler in complex notation. We have

$$a = Ae^{i\delta} \quad \text{(the complex amplitude)},$$

$$x = A\cos(\omega t + \delta) = \operatorname{Re} Ae^{i(\omega t+\delta)} = \operatorname{Re} a e^{i\omega t}, \qquad [5.1]$$

and the superposition principle,

$$a = a_1 + a_2, \qquad [5.2]$$

for the superposition of two oscillations. A geometrical illustration of superposition is given in Fig. 5.1. As long

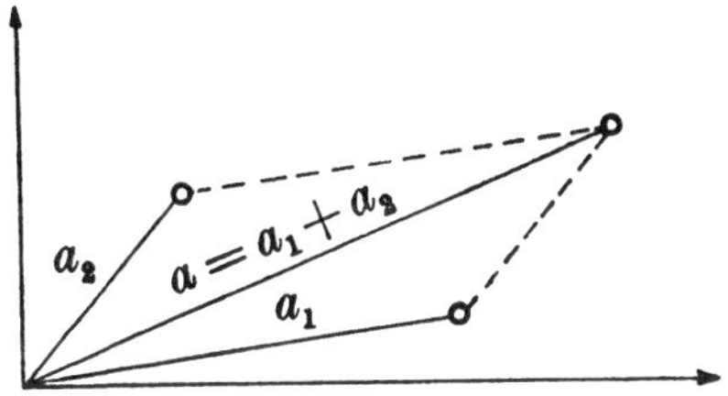

Figure 5.1

as one deals only with linear operations, everything proceeds very simply when written in complex form. However, care must be exercised when using quadratic expressions; thus, in general, the intensity I is

$$I \sim (\operatorname{Re} ae^{i\omega t})^2 = (\operatorname{Re} u)^2 = x^2, \qquad \text{with} \qquad u = ae^{i\omega t}.$$

Then,

$$I \sim \tfrac{1}{2}(u^* + u)^2 = u^*u + \tfrac{1}{2}(u^2 + u^{*2}) = |a|^2 + \operatorname{Re} a^2 e^{2i\omega t},$$

$$I \sim |a|^2\{1 + \cos(2\omega t + 2\delta)\}.$$

Within the bracket the first term is constant, while the second term is oscillatory. This oscillating term in the intensity cannot be simply ignored, except for the case when ω is large and a time-averaged value of I is considered.

Vectorial oscillations. Here, u is a vector instead of a scalar:

$$\boldsymbol{x} = \operatorname{Re} \boldsymbol{a} e^{i\omega t} = \operatorname{Re} \boldsymbol{u} ,$$

and $\boldsymbol{a} = \{a_1, a_2, a_3\}$ is a complex vector, i.e., the a_i are complex. The trajectory of a point performing such an oscillation lies in a fixed plane and is an ellipse (which may have the degenerate forms of a circle or a straight line).

Proof. a. The trajectory lies in a plane.

$$\boldsymbol{v} = \frac{\mathrm{d}\boldsymbol{x}}{\mathrm{d}t} = \operatorname{Re} i\omega \boldsymbol{a} e^{i\omega t} ,$$

$$\boldsymbol{x} \times \boldsymbol{v} = \operatorname{Re} \tfrac{1}{2} (\boldsymbol{a} e^{i\omega t} + \boldsymbol{a}^* e^{-i\omega t}) \times i\omega \boldsymbol{a} e^{i\omega t} = \operatorname{Re} \frac{\omega}{2} i\boldsymbol{a}^* \times \boldsymbol{a} ,$$

since $\boldsymbol{a} \times \boldsymbol{a} = 0$. However, $i\boldsymbol{a}^* \times \boldsymbol{a} = \boldsymbol{c}$ is real, since $\boldsymbol{c}^* = \boldsymbol{c}$. Thus $\boldsymbol{x} \times \boldsymbol{v} = \omega \boldsymbol{c}/2$ is constant, so that as a consequence the point lies in a plane perpendicular to $\boldsymbol{c}$ throughout the entire motion.

The trajectory is an ellipse. The plane of the motion is chosen as the X-Y plane. Then,

$$X = \tfrac{1}{2} (a_1 e^{i\omega t} + a_1^* e^{-i\omega t}) ,$$

$$Y = \tfrac{1}{2} (a_2 e^{i\omega t} + a_2^* e^{-i\omega t}) .$$

With

$$a = a_1 + i a_2 , \qquad a^* = a_1^* - i a_2^* ,$$

$$b = a_1 - i a_2 , \qquad b^* = a_1^* + i a_2^* ,$$

it follows that

$$X + iY = \tfrac{1}{2} (a e^{i\omega t} + b^* e^{-i\omega t}) ,$$

$$X - iY = \tfrac{1}{2} (b e^{i\omega t} + a^* e^{-i\omega t}) .$$

We now rotate the coordinate system (X, Y) into (ξ, η):

$$\xi + i\eta = (X + iY)e^{i\alpha} = \tfrac{1}{2}(a' e^{i\omega t} + b'^* e^{-i\omega t}) ,$$
$$\xi - i\eta = (X - iY)e^{-i\alpha} = \tfrac{1}{2}(b' e^{i\omega t} + a'^* e^{-i\omega t}) ,$$

where

$$a' = ae^{i\alpha} , \qquad a'^* = a^* e^{-i\alpha} ,$$
$$b' = be^{-i\alpha} , \qquad b'^* = b^* e^{i\alpha} ,$$

and determine the angle of rotation α in such a manner that the quotient a'/b' becomes real and positive. (The coordinates corresponding to this rotation are called normal coordinates.)

The special case in which $b = 0$ yields $X + iY = \frac{1}{2}ae^{i\omega t}$, which is clearly circular motion, and the same is true for $a = 0$ [A-2]. Excluding this special case, it follows that

$$b'/a' = e^{-2i\alpha} b/a = \varrho \qquad \text{(real and positive)} ,$$
$$b/a = \varrho e^{2i\alpha} ,$$

whereupon 2α is determined to within modulo 2π, and α, therefore, to within modulo π. There follows

$$\xi + i\eta = \tfrac{1}{2}(a' e^{i\omega t} + \varrho a'^* e^{-i\omega t}) ,$$
$$\xi - i\eta = \tfrac{1}{2}(\varrho a' e^{i\omega t} + a'^* e^{-i\omega t}) ,$$

and, with $a' = 2Ce^{i\delta}$,

$$\left.\begin{aligned} \xi &= C(1+\varrho)\cos(\omega t + \delta) \\ \eta &= C(1-\varrho)\sin(\omega t + \delta) \end{aligned}\right\} ,$$

which are the equations of an ellipse, since eliminating the time factors yields

$$\frac{\xi^2}{[C(1+\varrho)]^2} + \frac{\eta^2}{[C(1-\varrho)]^2} = 1 .$$

An oscillation along a straight line follows for $\varrho = 1$. Q.E.D.

Fourier's theorem permits the representation of periodic

processes as a superposition of harmonic processes:

$$\left.\begin{aligned} f(t+T) &= f(t) \\ f(t) &= \sum_{-\infty}^{+\infty} a_n e^{in\omega t} \\ a_n &= \frac{\omega}{2\pi}\int_{t_0}^{t_0+T} f(t)\, e^{-in\omega t}\, dt \end{aligned}\right\}, \qquad [5.3]$$

where $\omega = 2\pi/T$ is the circular frequency. For the formulas of Eq. [5.3] to be valid, f need not be continuous but only to be of bounded variation.

If f is real, then there follows

$$a_{-n} = a_n^* \,.$$

Furthermore, because of the orthogonality of the factors $e^{in\omega t}$, we have

$$\frac{1}{T}\int_{t_0}^{t_0+T} \Big| f(t) - \sum_{n=-N}^{+N} a_n e^{in\omega t} \Big|^2 dt = \frac{1}{T}\int | f(t) |^2 dt - 2\sum_{-N}^{N} a_n a_n^* + \sum_{-N}^{N} a_n a_n^*$$

$$= \frac{1}{T}\int | f(t) |^2 dt - \sum_{-N}^{N} | a_n |^2$$

for the coefficients a_n. From this follows Bessel's inequality

$$\frac{1}{T}\int | f(t) |^2 dt \geqslant \sum_{-N}^{+N} | a_n |^2 , \qquad [5.4]$$

which for $N \to \infty$ becomes the completeness relation

$$\frac{1}{T}\int | f(t) |^2 dt = \sum_{-\infty}^{+\infty} | a_n |^2 . \qquad [5.5]$$

The Fourier formulas can also be generalized to include nonperiodic motions by formally carrying out the following limiting process:

$$\omega \to 0\,, \qquad n \to \infty\,, \qquad n\omega \to \nu \quad \text{(fixed)}\,, \qquad T \to \infty\,,$$

$$\omega = \Delta\nu\,, \qquad \omega \sum \to \int \mathrm{d}\nu\,, \qquad a_n = \omega A(\nu)\,.$$

Then,

$$\left.\begin{aligned} A(\nu) &= \frac{1}{2\pi}\int_{-\infty}^{+\infty} f(t)\,e^{-i\nu t}\,\mathrm{d}t \\ f(t) &= \int_{-\infty}^{+\infty} A(\nu)\,e^{i\nu t}\,\mathrm{d}\nu \end{aligned}\right\}. \qquad [5.6]$$

The completeness relation becomes Parseval's formula,

$$\int_{-\infty}^{+\infty} |\,A(\nu)\,|^2\,\mathrm{d}\nu = \frac{1}{2\pi}\int_{-\infty}^{+\infty} |\,f(t)\,|^2\,\mathrm{d}t\,. \qquad [5.7]$$

For this limiting process to be permissible, all integrals must of course exist, and the functions must vanish "sufficiently fast" at infinity.

b. Waves

A plane wave progressing in the x direction is represented by

$$u = A\cos(kx - \omega t + \delta)\,,$$

for which

$$u(x, t) = u(x, t + \tau)\,, \qquad [5.8]$$

where

$$\left.\begin{aligned} \omega &= 2\pi\nu = \frac{2\pi}{\tau}\,, \qquad \nu = \frac{1}{\tau} \\ k &= 2\pi\mu = \frac{2\pi}{\lambda}\,, \qquad \mu = \frac{1}{\lambda} \end{aligned}\right\}. \qquad [5.9]$$

For

$$x = vt + \text{constant} = \frac{\omega}{k}\,t + \text{constant}\,,$$

the phase of the cosine,

$$\varphi \equiv kx - \omega t + \delta ,$$

remains constant. Thus,

$$v = \frac{\omega}{k} \qquad [5.10]$$

is the *phase velocity* of the wave.

In general, there exists a relation between ω and k:

$$\omega = f(k) , \qquad \text{dispersion law} . \qquad [5.11]$$

The special case where $\omega/k = c$ holds for light in vacuum.

In the three-dimensional case, we use the vectors $\boldsymbol{x}$ and $\boldsymbol{k}$:

$$u = A \cos (\boldsymbol{k} \cdot \boldsymbol{x} - \omega t + \delta) . \qquad [5.8']$$

Thus, $\boldsymbol{k}$ is perpendicular to the surfaces of constant phase. The wave normal $\boldsymbol{n}$ is related to $\boldsymbol{k}$ by

$$\boldsymbol{k} = \boldsymbol{n} \frac{2\pi}{\lambda} , \qquad \text{with } \boldsymbol{n}^2 = 1 . \qquad [5.9']$$

The dispersion law becomes

$$\omega = f(k_1, k_2, k_3) = f(\boldsymbol{k}) . \qquad [5.11']$$

If the medium is isotropic, then we have simply

$$\omega = f(|\boldsymbol{k}|) .$$

The *principle of linear superposition* is a general property of the waves considered here. It states that *if u_1 and u_2 are two possible wave states of a medium, then so is $c_1 u_1 + c_2 u_2$.* In order that u_1 and u_2 be two *possible* states, the dispersion law must be satisfied.

The principle of linear superposition is of very far-reaching significance in wave theory, and it is also well satisfied in optics [A-3]. For mechanical waves, however, it is valid only for sufficiently small amplitudes. (See, for example, a work of Riemann on explosion waves in air.) Because of the superposition principle, Fourier's theorem permits an arbitrary wave state in a medium to be represented as follows:

$$\left.\begin{aligned} u &= \iiint A(k_1, k_2, k_3)\cos[\boldsymbol{k}\cdot\boldsymbol{x} - \omega t + \delta(\boldsymbol{k})]\,\mathrm{d}^3k \\ &= \mathrm{Re}\iiint a(\boldsymbol{k})\,e^{i(\boldsymbol{k}\cdot\boldsymbol{x}-\omega t)}\,\mathrm{d}^3k \end{aligned}\right\}, \qquad [5.12]$$

where $a = Ae^{i\delta}$. In other words, if any dispersion law of the form [5.11′] is given, and if (for optics) it is demanded that the principle of linear superposition be valid, then the general solution of the problem is given by formula [5.12].

With formula [5.12], the initial-value problem

$$u(x, 0) = u_0(x)\,,$$

$$\left(\frac{\partial u}{\partial t}\right)_{t=0} = \dot{u}_0(x)\,,$$

is solvable; this is because

$$u_0 = \mathrm{Re}\iiint a(\boldsymbol{k})\,e^{i\boldsymbol{k}\cdot\boldsymbol{x}}\,\mathrm{d}^3k\,,$$

$$\dot{u}_0 = \mathrm{Re}\iiint (-i)\omega\, a(\boldsymbol{k})\,e^{i\boldsymbol{k}\cdot\boldsymbol{x}}\,\mathrm{d}^3k = \mathrm{Im}\iiint \omega(\boldsymbol{k})\,a(\boldsymbol{k})\,e^{i\boldsymbol{k}\cdot\boldsymbol{x}}\,\mathrm{d}^3k\,,$$

from which the complex quantity a is determined by means of the Fourier inversion formula:

$$\frac{1}{2}[a(k) + a^*(-k)] = \left(\frac{1}{2\pi}\right)^3\iiint u_0(x)\,e^{-i\boldsymbol{k}\cdot\boldsymbol{x}}\,\mathrm{d}^3x\,,$$

$$(-i)\frac{1}{2}[\omega(k)\,a(k) - \omega\,(-k)\,a^*(-k)] = \left(\frac{1}{2\pi}\right)^3\iiint \dot{u}_0(x)\,e^{-i\boldsymbol{k}\cdot\boldsymbol{x}}\,\mathrm{d}^3x\,.$$

Group velocity. We consider a wave packet, that is, a superposition of many waves of the form given by Eq. [5.12]:

$$u(x, t) = \mathrm{Re}\iiint a(\boldsymbol{k})\,e^{i(\boldsymbol{k}\cdot\boldsymbol{x}-\omega t)}\,\mathrm{d}^3k = \mathrm{Re}\,w\,.$$

Here, it is assumed that $a(\boldsymbol{k})$ is significantly different from zero only in some small region around $\boldsymbol{k}$. The center of the packet is denoted by

$$\bar{x}_i(t) = \frac{\int x_i u^2 \, \mathrm{d}V}{\int u^2 \, \mathrm{d}V} . \qquad [5.13]$$

We will see that the energy is proportional to u^2, so that in Eq. [5.13] we have the "center" of the energy. We have

$$u(x, t) = \tfrac{1}{2}(w + w^*) \qquad \text{with} \qquad w = \iiint a(\boldsymbol{k}) \, e^{i(\boldsymbol{k}\cdot\boldsymbol{x} - \omega t)} \, \mathrm{d}^3 k ,$$

so that

$$e^{-i\omega t} a(\boldsymbol{k}) = \frac{1}{(2\pi)^3} \iiint w \, e^{-i\boldsymbol{k}\cdot\boldsymbol{x}} \, \mathrm{d}^3 x , \qquad [5.14]$$

and

$$u^2 = \tfrac{1}{4}(w^2 + w^{*2}) + \tfrac{1}{2} w w^* .$$

Thus, u^2 contains two types of terms, constants and oscillating terms of twice the frequency. The oscillating terms are neglected (time averaging), after which it follows that

$$\bar{x}_i(t) = \frac{\int x_i w w^* \, \mathrm{d}V}{\int w w^* \, \mathrm{d}V} . \qquad [5.14']$$

The numerator becomes

$$\begin{aligned} \int x_i w w^* \, \mathrm{d}V &= \int \mathrm{d}^3 x \, w^* \int a(\boldsymbol{k}) \, \frac{e^{-i\omega t}}{i} \, \frac{\partial}{\partial k_i} (e^{i\boldsymbol{k}\cdot\boldsymbol{x}}) \, \mathrm{d}^3 k \\ &= \int \mathrm{d}^3 x \, w^* \int \left(i \frac{\partial a}{\partial k_i} + t \frac{\partial \omega}{\partial k_i} a \right) e^{i(\boldsymbol{k}\cdot\boldsymbol{x} - \omega t)} \, \mathrm{d}^3 k \\ &= (2\pi)^3 \left(\int a^* i \frac{\partial a}{\partial k_i} \, \mathrm{d}^3 k + t \int \frac{\partial \omega}{\partial k_i} a^* a \, \mathrm{d}^3 k \right) , \end{aligned}$$

where we have integrated by parts and used Eq. [5.14]. With Parseval's formula (Eq. [5.7]), the denominator becomes

$$\int w w^* \, \mathrm{d}^3 x = (2\pi)^3 \int a a^* \, \mathrm{d}^3 k ,$$

whereupon

$$\bar{x}_i(t) = \frac{\int a^* i\,(\partial a/\partial k_i)\,\mathrm{d}^3k}{\int a^* a\,\mathrm{d}^3k} + \frac{\int (\partial\omega/\partial k_i)\,a^* a\,\mathrm{d}^3k}{\int a^* a\,\mathrm{d}^3k}\,t\,.$$

Thus,

$$\frac{\partial \bar{x}_i}{\partial t} = \overline{\left(\frac{\partial\omega}{\partial k_i}\right)} \qquad [5.15]$$

is the velocity of propagation of the center of the wave packet.

The *group velocity* is defined by

$$\boldsymbol{U} = \frac{\partial\omega}{\partial \boldsymbol{k}}\,. \qquad [5.16]$$

In optics, $\boldsymbol{U}$ has the direction of the ray in geometrical optics. In deriving the group velocity, we arbitrarily took u^2 as a measure of the energy. Exactly the same result is obtained if one takes instead any quadratic form $\varphi(\omega)u^2$ (for example $(\partial u/\partial t)^2$, since $\partial/\partial t \sim \omega$).

Isotropic case. We consider a wave packet, Eq. [5.12]. For the isotropic case we must have

$$\omega = \omega(k)\,, \qquad k = |\boldsymbol{k}|\,.$$

If the packet is sufficiently small, then

$$\bar{v} \simeq v = \frac{\omega}{k}\,, \qquad U = v + k\frac{\partial v}{\partial k} = v - \lambda\frac{\partial v}{\partial \lambda}\,.$$

In the isotropic case, for any arbitrary individual wave of the wave packet, we can always set

$$\nabla^2 u + k^2 u = 0\,, \qquad [5.17]$$

since for isotropy the relation

$$\nabla^2 = -k^2$$

holds for the integrands in Eq. [5.12]. On the one hand,

Eq. [5.17] is special in that it is limited to a process which is sinusoidal in time; on the other hand, however, it is general, since no special dispersion law has been assumed. It is also an expression of the superposition principle, but only in so far as it refers to wave states of the same frequency.

Special case. Let the medium be free of dispersion: $v = U = c$ for all waves of the packet. It is then true for *all* waves of the packet that

$$c = \frac{\omega}{k} = \text{constant}. \qquad [5.17']$$

From formula [5.12] it follows that for each wave of the packet

$$\frac{\partial^2}{\partial t^2} = -\omega^2 = -c^2 k^2. \qquad [5.18]$$

Thus, upon substitution into Eq. [5.17], the *wave equation*,

$$\nabla^2 u - \frac{1}{c^2}\frac{\partial^2 u}{\partial t^2} = 0, \qquad [5.18']$$

follows.

From Eq. [5.12], which is the general expression of the principle of linear superposition for wave kinematics, we have derived a differential equation by taking as a basis the quite special dispersion law of Eq. [5.17′]. In this special case—*and only in this case*—Eqs. [5.12] and [5.18] are equivalent. The linearity of the superposition principle [5.12] is reflected in the linearity of the differential equation [5.18′].

One-dimensional case. We consider a single position coordinate x for the special case where Eq. [5.18′] is valid,

$$\frac{\partial^2 u}{\partial x^2} - \frac{1}{c^2}\frac{\partial^2 u}{\partial t^2} = 0. \qquad [5.18'']$$

According to Eq. [5.12], the solution is

$$2u = \int_{-\infty}^{+\infty} a(k)\, e^{i(kx - c|k|t)}\, \mathrm{d}k + \text{complex conjugate}$$

$$= \int_{0}^{\infty} a(k)\, e^{ik(x-ct)}\, \mathrm{d}k + \int_{0}^{\infty} a(-k)\, e^{-ik(x+ct)}\, \mathrm{d}k + \text{complex conjugate}\,.$$

Since $a(-k)$ is completely independent of $a(k)$,

$$2u = f(x - ct) + g(x + ct) \tag{5.19}$$

follows as the most general solution. Equation [5.19] can also be derived without Fourier integrals by the direct solution of the wave equation [5.18′] for this case.

In the same way, for spherical waves in an isotropic, dispersion-free medium, it follows that

$$ru = f(r - ct) + g(r + ct)\,. \tag{5.19′}$$

To show this, one need only to substitute $v = ru$ in the wave equation [5.18′], and Eq. [5.18″] (with $x = r$) follows for v. If there is to be no singularity at $r = 0$, then one must require

$$ru = f(r - ct) - f(-r - ct)\,.$$

6. REFRACTION, REFLECTION, INTERFERENCE

We now apply the general wave theory treated in Section 5 to optics. For this to be at all permissible, it must be shown that the ray theory treated in Chapter 1 is contained within the wave theory, or that it is at least a certain limiting case of the latter. This will now be shown for the fundamental laws of refraction and reflection.

Let a bounding surface F be given which separates regions characterized by k, ω and k', ω'. Our considerations will apply to a plane wave in an isotropic medium.

We require continuity of the phase at the boundary F. This requires

$$\omega = \omega' . \tag{6.1}$$

Furthermore, $\boldsymbol{k}\cdot\boldsymbol{x} = \boldsymbol{k}'\cdot\boldsymbol{x}'$ for $\boldsymbol{x} = \boldsymbol{x}'$ on F. Thus, $(\boldsymbol{k}-\boldsymbol{k}')\cdot\boldsymbol{x} = 0$ for $\boldsymbol{x}$ on F. Hence, $(\boldsymbol{k}-\boldsymbol{k}')$ must be parallel to the surface normal $\boldsymbol{N}$. This, however, is the law of refraction:

$$(\boldsymbol{k}-\boldsymbol{k}')\times\boldsymbol{N} = \boldsymbol{k}\times\boldsymbol{N} - \boldsymbol{k}'\times\boldsymbol{N} = 0 ,$$

so that

$$k\sin\vartheta = k'\sin\vartheta' . \tag{6.1'}$$

Comparing this with [1.1], we see that

$$n/n' = k/k' = \lambda'/\lambda . \tag{6.2}$$

We have remarked that in the case of isotropy, the phase velocity is $v = \omega/k$. Substituting this into Eq. [6.2] and then using [6.1] gives

$$\frac{n}{n'} = \frac{v'}{v} . \tag{6.3}$$

The absence of dispersion in vacuum is an experimental fact; consequently, in our theory a universal velocity c must be assumed for light in vacuum. Furthermore, if $n = 1$ in vacuum by definition, then it follows that for an arbitrary medium with index of refraction n (with respect to vacuum) we have

$$v = \frac{c}{n} . \tag{6.3'}$$

This is the relation between refractive index and phase velocity which can be verified by direct experiment. The numerical value of c must be determined experimentally; it is not predicted by the theory.

Continuity of the wave field at the bounding surface F is also required for the reflected ray. Just as before, it follows that $\omega = \omega'$ and $(\boldsymbol{k} - \boldsymbol{k}') \cdot \boldsymbol{x} = 0$ for $\boldsymbol{x}$ on F. However, from the above we now have $n = n'$ (the same medium), so that $|\boldsymbol{k}| = |\boldsymbol{k}'|$. Thus, $(\boldsymbol{k} - \boldsymbol{k}')$ is perpendicular to F, or $(\boldsymbol{t} - \boldsymbol{t}')$ is perpendicular to F, where $|\boldsymbol{t}| = 1$. This is the law of reflection.

In order to show more generally that the ray theory is contained in, or at least represents a limiting case of, the wave theory, let us now assume an inhomogeneous isotropic medium. If processes harmonic in time are assumed, then Eq. [5.17] is valid for such a medium:

$$\nabla^2 u + k^2 u = 0 \,, \qquad [5.17]$$

where k is a function of position. Introducing the refractive index in accordance with Eq. [6.2], we obtain

$$\nabla^2 u + k_0^2 n^2 u = 0 \,, \qquad [6.4]$$

where k_0 is the value of k corresponding to vacuum. For u we assume

$$u = \operatorname{Re} e^{i(k_0 S - \omega t)} \,, \qquad [6.5]$$

where $S = S(\boldsymbol{x})$ is complex. In addition, $u_1 = e^{ik_0 S}$ must be a solution of Eq. [6.4]. We have

$$\frac{\partial u_1}{\partial x} = ik_0 \frac{\partial S}{\partial x} u_1 \,,$$

$$\frac{\partial^2 u_1}{\partial x^2} = \left\{ \left(ik_0 \frac{\partial S}{\partial x} \right)^2 + ik_0 \frac{\partial^2 S}{\partial x^2} \right\} u_1 \,.$$

Upon substitution into Eq. [6.4] and simplification, the equation

$$-(\nabla S)^2 + i \frac{1}{k_0} \nabla^2 S + n^2 = 0 \qquad [6.6]$$

follows for S.

In geometrical optics there is a similar equation, namely,

$$-(\nabla S_0)^2 + n^2 = 0$$

(see Eq. [2.2]). Furthermore, in geometrical optics there

are also the relations

$$\nabla S_0 = \boldsymbol{p} = n\boldsymbol{t}, \qquad \nabla^2 S_0 = \operatorname{div} \boldsymbol{p}.$$

In order that Eq. [6.6] be equivalent to the equation from geometrical optics, we must require that

$$\frac{1}{k_0}(\operatorname{div} \boldsymbol{p}) \ll p^2. \tag{6.7}$$

In the one-dimensional case, this yields

$$\frac{1}{k_0}\left|\frac{\mathrm{d}p}{\mathrm{d}x}\right| \ll p^2, \qquad \left|\frac{\mathrm{d}n}{\mathrm{d}x}\right| \ll k_0 n^2,$$

$$\left|\frac{\lambda}{2\pi}\frac{\mathrm{d}n}{\mathrm{d}x}\right| \ll n. \tag{6.8}$$

Thus, the change in n over the distance of a wavelength must be small—in that case wave theory and ray theory give the same results. Condition [6.8] is certainly satisfied if the wavelength λ is very small. Thus, we can expand S in powers of λ, or of $1/k$, to obtain directly

$$S = S_0 + \frac{1}{k_0} S_1 + \dots + \frac{1}{k_0^n} S_n + \dots,$$

where S_0 is a solution of the equation from geometrical optics (the first approximation in our expansion, if all terms with powers of $1/k_0$ or higher are neglected).

Accordingly, it is seen that if $\lambda \to 0$ is assumed, then for the case of isotropic media our wave theory describes phenomena in the same way as geometrical optics does. The question now arises whether the wave theory is a better approximation to nature than is the ray theory for cases where the finiteness of λ is of experimental significance.

Interference phenomena show that we are on the right path when using the wave picture for light. We set the light intensity introduced in Section 4 (the illuminance, for example) proportional to the square of the wave function. Then, in our new theory, in contrast to geometrical optics, the intensities are not additive; this is, indeed, observed to be the case. For example, let two (harmonic) waves be superimposed at a given point. If

$$u_1 = \mathrm{Re}\, a_1 e^{-i\omega t} ,$$
$$u_2 = \mathrm{Re}\, a_2 e^{-i\omega t}$$

are the individual excitations of the two waves, then

$$u_1 + u_2 = \mathrm{Re}\,(a_1 + a_2)\, e^{-i\omega t}$$

is the total excitation at the point under consideration. Thus, there follows

$$I_1 \propto |a_1|^2 , \qquad I_2 \propto |a_2|^2 ,$$
$$I \propto |a_1 + a_2|^2 = (a_1 + a_2)(a_1^* + a_2^*) ,$$

so that

$$I = I_1 + I_2 + 2\sqrt{I_1 I_2} \cos\delta , \qquad [6.9]$$

where δ represents the difference of the arguments of a_1 and a_2. If this phase difference is well defined, then one speaks of coherent wave trains. In this case, the nonadditivity of the intensities is evident. If, however, we average over all possible δ, that is, light from separate sources, then it again follows that $I = I_1 + I_2$.

Coherent light can be produced in Fresnel's mirror experiment. In Fig. 6.1, L is a light source which produces

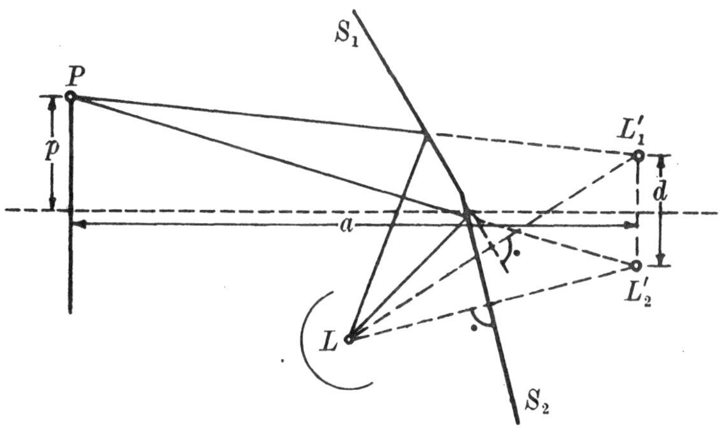

Figure 6.1

the virtual sources L_1' and L_2' by reflections from mirrors S_1 and S_2, respectively. We calculate the intensity at the point P:

$$r_1 = PL_1' = PS_1L\,, \qquad r_1^2 = a^2 + (\tfrac{1}{2}d - p)^2\,,$$

$$r_2 = PL_2' = PS_2L\,, \qquad r_2^2 = a^2 + (\tfrac{1}{2}d + p)^2\,,$$

$$r_2^2 - r_1^2 = 2dp\,,$$

$$r_1 + r_2 \simeq 2a\,, \qquad \text{so that} \qquad r_1 - r_2 \simeq \frac{dp}{a}\,.$$

Thus, there follows

$$\delta = \frac{2\pi}{\lambda}\frac{dp}{a}\,,$$

which with Eq. [6.9] gives

$$I = 2I_0(1 + \cos\delta) = 4I_0\cos^2\delta/2\,.$$

From this we infer that intensity maxima are obtained for $\delta = 2\pi n$ (n integral), that is, for $p = an\lambda/d$. In between, for $p = (\lambda a/2d)(2n+1)$, are points at which the intensity is zero.

An additional experiment for producing interference phenomena is the combining of rays which have the same

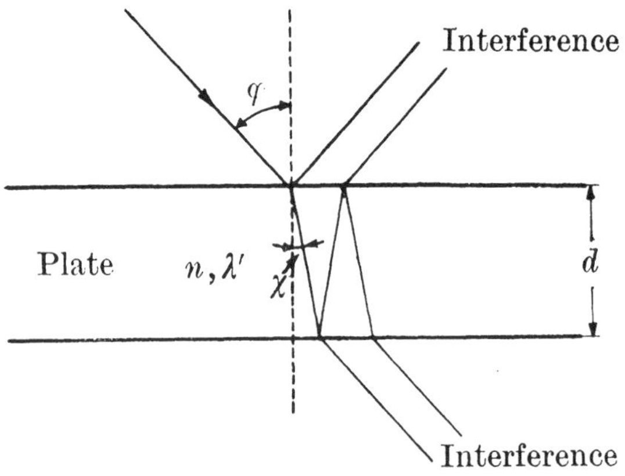

Figure 6.2

inclination after having been reflected from parallel surfaces (Fig. 6.2). A calculation yields the following result: The condition

$$2d \cos \chi = m\lambda'$$

implies maxima below (transmission) and minima above (reflection), while for

$$2d \cos \chi = (m + \tfrac{1}{2})\lambda'$$

we have minima below and maxima above.

Newton's rings are another interference phenomenon. This experiment demonstrates that the wavelength decreases as the color of light changes from red to violet.

7. THEORY OF DIFFRACTION

a. General solution of Kirchhoff

According to geometrical optics, small openings should always cast sharp shadows. Experimentally, however, the familiar phenomena designated as diffraction are observed. Thus, we attempt to explain these phenomena on the basis of the wave concept developed here.

So as not to have to calculate with the general formula [5.12], we assume that the entire wave field is harmonic in time, with the time factor $e^{-i\omega t}$. It is clear (because of continuity) that if at *one* point of the wave field we assume an excitation of the form $e^{-i\omega t}$ unbounded in time, then this must also be done for all other points, and with *the same* ω. In this way, a mathematically correct formulation is obtained; in nature, however, this idealization is only approximately satisfied, since in reality one must always deal with wave packets (of finite duration).

With this assumption, upon requiring an isotropic medium, Eq. [5.17] holds:

$$\nabla^2 u + k^2 u = 0 .$$

In diffraction phenomena, the screens have the significance of boundary conditions for u and $\partial u/\partial n$. Equation [5.17] is to be solved for these boundary conditions. It should be mentioned at the outset that this mathematical problem is exactly soluble only for very special cases. For the most part, we must restrict ourselves to approximations. Here, we will treat the approximation of Kirchhoff.

We have Green's theorem for a closed surface:

$$\int_{V_Q} (v\,\nabla^2 u - u\,\nabla^2 v)\,\mathrm{d}V = \oint_{F_Q} \left(v\frac{\partial u}{\partial n} - u\frac{\partial v}{\partial n}\right)\mathrm{d}f\,.$$

In this formula we substitute a solution of [5.17] for u, and for v we set

$$v = \frac{e^{ikr_{PQ}}}{r_{PQ}}\,.$$

(The light source L is outside the surface F.) The point P is an arbitrary field point and Q is the point which varies over the region of integration. Thus,

$$\nabla^2 v + k^2 v = 0\,,$$

so that for $Q \neq P$

$$u\,\nabla^2 v - v\,\nabla^2 u = 0\,.$$

If P is within the region of integration, then it must be excluded by means of a small sphere:

$$\oint_{F} \left(v\frac{\partial u}{\partial n} - u\frac{\partial v}{\partial n}\right)\mathrm{d}f = 0 \qquad \text{for } P \text{ outside } V,$$

$$\oint_{F+\text{sphere}} \left(v\frac{\partial u}{\partial n} - u\frac{\partial v}{\partial n}\right)\mathrm{d}f = 0 \qquad \text{for } P \text{ inside } V.$$

In calculating the contribution of the small sphere, we note that $\partial/\partial n = -\partial/\partial r$ on the sphere. For $r \to 0$, only the term

$$-\frac{\partial v}{\partial n} = \frac{\partial v}{\partial r} = \left[\frac{ik}{r} - \frac{1}{r^2}\right] e^{ikr}$$

is significant. On the sphere, $\mathrm{d}f = r^2 \mathrm{d}\Omega$, and there follows

$$\oint_{\text{sphere}} \mathrm{d}f \left(v \frac{\partial u}{\partial n} - u \frac{\partial v}{\partial n} \right) = - \oint \mathrm{d}\Omega\, u_P = - 4\pi u_P\,.$$

Thus

$$\oint_F \left(v \frac{\partial u}{\partial n} - u \frac{\partial v}{\partial n} \right) \mathrm{d}f = \begin{cases} 4\pi u_P & \text{for } P \text{ inside the surface}\,, \\ 0 & \text{for } P \text{ outside the surface}\,. \end{cases} \qquad [7.1]$$

We now let a part of the closed surface go to infinity. Furthermore, we require that the solution of Eq. [5.17] go asymptotically as

$$u \sim \frac{e^{ikr}}{r}$$

at infinity, and that there be no contributions of the form e^{-ikr}. This is a boundary condition at infinity, the radiation condition. Thus, physically, only outgoing waves are allowed. With this, however, it follows for the integrands that

$$\left(v \frac{\partial u}{\partial n} - u \frac{\partial v}{\partial n} \right) r^2 \,\mathrm{d}\Omega \to 0 \qquad \text{for } r \to \infty\,.$$

Thus, we need only to integrate over the finite part of the surface. There follows

$$\int_F \left(v \frac{\partial u}{\partial n} - u \frac{\partial v}{\partial n} \right) \mathrm{d}f = \begin{cases} 4\pi u_P & \text{if } P \text{ and } L \text{ are on different sides of } F, \\ 0 & \text{if } P \text{ and } L \text{ are on the same side of } F. \end{cases} \qquad [7.2]$$

Here, L denotes the light source. Formula [7.2] is exact if one already has a solution of the problem. In actuality, however, one usually knows neither u nor $\partial u/\partial n$ on such a surface; also, these quantities cannot be given in advance.

Kirchhoff's method. The surface is divided into screens and apertures: $F = S + A$. Then, Kirchhoff assumed that

$$\text{in } A: \quad u = u^0 , \quad \frac{\partial u}{\partial n} = \frac{\partial u^0}{\partial n} \quad \text{(undisturbed propagation);}$$

$$\text{on } S: \quad u = 0 , \quad \frac{\partial u}{\partial n} = 0 \quad \text{(vanishing light excitation).}$$

The quantity u^0 is to be understood as the form of u appropriate for undisturbed propagation of light:

$$u^0(Q) \equiv u^0_Q = \text{constant} \frac{e^{ikr_{LQ}}}{r_{LQ}} ,$$

where L denotes the light source.

If Kirchhoff's hypothesis is substituted into Eq. [7.2], then there follows

$$u_P = \frac{1}{4\pi} \int_A \left(v \frac{\partial u^0}{\partial n} - u^0 \frac{\partial v}{\partial n} \right) df$$

$$= u^0(P) - \frac{1}{4\pi} \int_S \left(v \frac{\partial u^0_Q}{\partial n} - u^0_Q \frac{\partial v}{\partial n} \right) df . \qquad [7.3]$$

This is Kirchhoff's solution of the wave equation. As already mentioned, this solution is not exact since the hypothesis is certainly not correct. If u^0_Q and $\partial u^0_Q / \partial n$ are calculated from Eq. [7.3], then the result is not at all what was put in. Also, the correct boundary conditions are not at all the assumed ones, since on the screens one can specify only *one* of the quantities u or $\partial u / \partial n$ (not both), while at the apertures neither can be given. With this hypothesis, Kirchhoff believed that he had found the solution for a completely "black" screen, but such a screen cannot even be defined.

Kirchhoff's solution is useful when the apertures are large as compared with the wavelength, and when the diffraction angle is not too large.

If the apertures and screens are interchanged, then *Babinet's theorem* follows from Kirchhoff's solution:

$$u_1(P) + u_2(P) = u^0(P) . \qquad [7.4]$$

(Here, u_1 is the solution for the screens at the positions S, and u_2 the solution for the apertures at S.) For example, for the case of assumed ideal imaging by a lens system of a light source L onto an objective screen (on such a screen $u_0 = 0$ everywhere except at a sharp image point L'), the light distribution ($\sim u^2$) is the same whether a given diaphragm is inserted, or whether another is used which has screens where the first had apertures and vice versa.

Improvement of Kirchhoff's solution. The function v can be chosen somewhat differently. We assume that the diffracting surface E is planar, and introduce the point P', which is the image of P in E. Then, we set

$$v = \frac{e^{ikr_{PQ}}}{r_{PQ}} - \frac{e^{ikr_{P'Q}}}{r_{P'Q}}$$

(where Q is the variable point over which we integrate). The new v has the properties

$$\left.\begin{aligned} v &= 0 \\ \frac{\partial v}{\partial n} &= 2\,\frac{\partial}{\partial n}\,\frac{e^{ikr_{PQ}}}{r_{PQ}} \end{aligned}\right\}$$

on E. The new solution $\bar{u}_P$ is

$$4\pi\bar{u}_P = 4\pi u_P^0 - \int_S (-u^0)\, 2\,\frac{\partial}{\partial n}\,\frac{e^{ikr}}{r}\,df . \qquad [7.5]$$

This is related to the old solution by

$$\bar{u} = u(P) - u(P') + u^0(P') .$$

Another possibility is to set

$$v = \frac{e^{ikr_{PQ}}}{r_{PQ}} + \frac{e^{ikr_{P'Q}}}{r_{P'Q}} ,$$

from which it follows that

$$\left.\begin{aligned} v &= 2\frac{e^{ikr}}{r} \\ \frac{\partial v}{\partial n} &= 0 \end{aligned}\right\}$$

on E,

$$4\pi\bar{\bar{u}}_P = 4\pi u_P^0 - 2\int \frac{\partial u^0}{\partial n}\frac{e^{ikr}}{r}\,df\,, \qquad [7.5']$$

and, correspondingly,

$$\bar{\bar{u}}_P = u(P) + u(P') - u^0(P')\,.$$

However, nothing essential is obtained in this way. The advantage of these solutions is that only *one* boundary condition need be specified, instead of the two that had to be specified before.

Consequences of Kirchhoff's solution. We now substitute the following functions u_Q^0 and v_Q:

$$v_Q = e^{ikr_0}/r_0\,,$$
$$u_Q^0 = Ae^{ikr_1}/r_1\,,$$
$$u^0(P) = Ae^{ikr}/r\,.$$

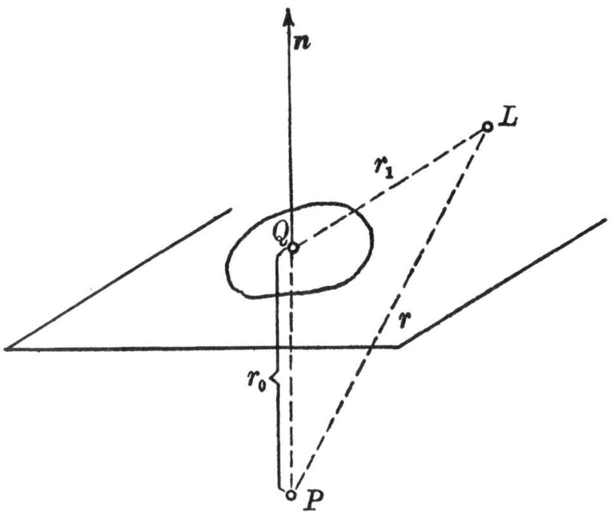

Figure 7.1

However,

$$\frac{\partial}{\partial n}\left(\frac{e^{ikr}}{r}\right) = \cos(\boldsymbol{n}, \boldsymbol{r}) \frac{\mathrm{d}}{\mathrm{d}r}\frac{e^{ikr}}{r} = \cos(\boldsymbol{n}, \boldsymbol{r})\, ik\, \frac{e^{ikr}}{r}\left(1-\frac{1}{ikr}\right).$$

Thus, from Eq. [7.3] there follows the formula

$$u_P = \frac{A}{4\pi}\int_A \frac{e^{ik(r_0+r_1)}}{r_0 r_1}\, ik$$

$$\times\left[\cos(\boldsymbol{n}, \boldsymbol{r}_1)\left(1-\frac{1}{ikr_1}\right) - \cos(\boldsymbol{n}, \boldsymbol{r}_0)\left(1-\frac{1}{ikr_0}\right)\right]\mathrm{d}f \qquad [7.6]$$

for the effect of a point light source.

We now make the following simplification:

$$kr_0 \gg 1\,, \quad kr_1 \gg 1 \qquad (\lambda \text{ very small compared with } r)\,.$$

In addition, we consider only small diffraction angles—this is done in consideration of the experimental fact that at small distances away from the geometrical shadow boundary, the light distribution is that given by geometrical optics. Now, because of the smallness of λ, the factor $e^{ik(r_0+r_1)}$ is a very rapidly oscillating function, while the other remaining factor, $\cos(\boldsymbol{n}, \boldsymbol{r}_1)-\cos(\boldsymbol{n}, \boldsymbol{r}_0)$, varies slowly (we have already taken $1/ikr\sim 0$). Thus, we can assume this remaining term to be a constant and replace it by $-2\cos\delta$, where δ is an appropriately chosen average angle. Then, from Eq. [7.6] there follows

$$u_P = -A\frac{ik}{2\pi}\cos\delta\int_A \frac{e^{ik(r_0+r_1)}}{r_0 r_1}\,\mathrm{d}f$$

$$= u^0 + A\frac{ik}{2\pi}\cos\delta\int_S \frac{e^{ik(r+r_1)}}{r r_1}\,\mathrm{d}f\,. \qquad [7.7]$$

If we assume that the diaphragm opening is such that a plane can be passed through it, then Eq. [7.7] can be expanded about a null ray. Let M be a point in the opening and $(\xi, \eta, 0)$ the cartesian coordinates of the integration point Q. (We have taken M as the origin of our co-

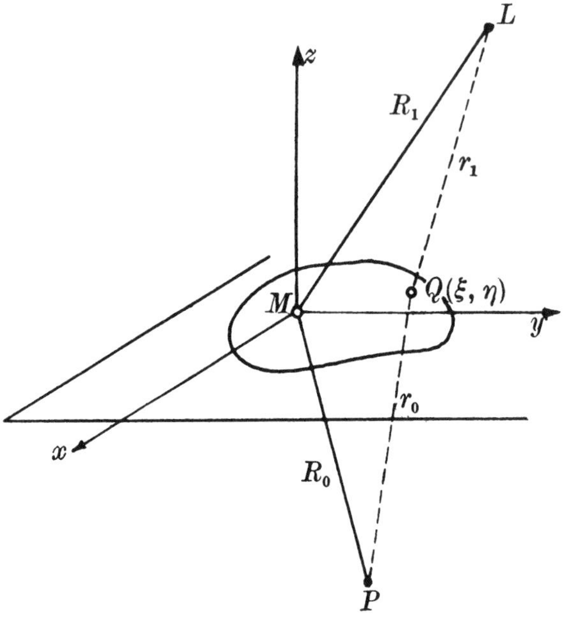

Figure 7.2

ordinate system, with the x-y plane in the plane of the diaphragm opening.) Then there follows

$$r_i^2 = (x_i - \xi)^2 + (y_i - \eta)^2 + z_i^2 , \qquad i = 0,1 ,$$

$$R_i^2 = x_i^2 + y_i^2 + z_i^2 .$$

Thus,

$$r_i^2 = R_i^2 - 2(x_i\xi + y_i\eta) + \xi^2 + \eta^2 .$$

However, we know that

$$\sqrt{1+\varepsilon^2} = 1 + \frac{\varepsilon^2}{2} - \frac{\varepsilon^4}{8} + \dots ,$$

so that

$$r_i = R_i - \frac{x_i\xi + y_i\eta}{R_i} + \frac{\xi^2 + \eta^2}{2R_i} - \frac{(x_i\xi + y_i\eta)^2}{2R_i^3} + \dots .$$

This substituted into Eq. [7.7] yields

$$u_P = -A\frac{ik}{2\pi}\cos\delta\frac{e^{ik(R_0+R_1)}}{R_0R_1}\iint e^{ik\Phi(\xi,\eta)}\,\mathrm{d}\xi\,\mathrm{d}\eta\,, \qquad [7.8]$$

where

$$\Phi(\xi,\eta) = -\frac{x_0\xi + y_0\eta}{R_0} + \frac{\xi^2 + \eta^2}{2R_0} - \frac{(x_0\xi + y_0\eta)^2}{2R_0^3} + \dots$$

$$-\frac{x_1\xi + y_1\eta}{R_1} + \frac{\xi^2 + \eta^2}{2R_1} - \frac{(x_1\xi + y_1\eta)^2}{2R_1^3} + \dots .$$

With formula [7.8], the light distribution for a diffraction phenomenon can be determined (in so far as Kirchhoff's solution is assumed to be correct) by direct calculation. The more terms considered in the expansion of the function Φ of Eq. [7.8], the more difficult it is to carry out the calculation. If only the linear terms in Φ are considered, then physically this implies that both the light source and the field point are infinitely far from the diffracting aperture. The quantities $x_i/R_i = \alpha_i$ and $y_i/R_i = \beta_i$ ($i = 0, 1$) become direction cosines, and Eq. [7.8] becomes

$$u_P = \text{constant}\cdot\int e^{-ik[\xi(\alpha_0+\alpha_1)+\eta(\beta_0+\beta_1)]}\,\mathrm{d}\xi\,\mathrm{d}\eta\,. \qquad [7.8']$$

In practice, one would of course not assume L and P to be very far away; instead, one would assume that parallel rays had been achieved with lenses. This case is known as *Fraunhofer diffraction.*

If the quadratic terms are also considered, then we have *Fresnel diffraction.* We will treat some cases of both types of diffraction.

b. Fraunhofer diffraction

1. *Diffraction by a slit.* Equation [7.8] is valid for the Fraunhofer case. If the slit (of width a) is infinitely long, then we have a plane problem:

$$\beta_0 + \beta_1 = 0\,,$$

$$\alpha_0 = \sin\psi = \cos\left(\frac{\pi}{2} - \psi\right),$$

$$\alpha_1 = -\sin\varphi = \cos\left(\frac{\pi}{2} + \varphi\right),$$

$$\mu = k(\alpha_0 + \alpha_1) = k(\sin\psi - \sin\varphi)\,.$$

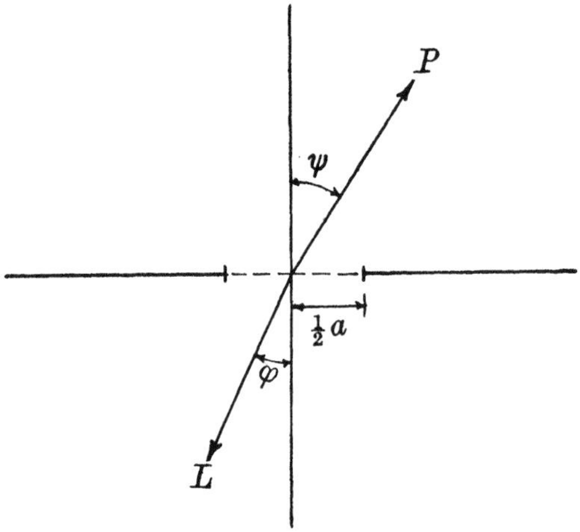

Figure 7.3

With these relations, Eq. [7.8] yields for the slit

$$u_P = \text{constant}\cdot\int_{-a/2}^{+a/2} e^{-i\mu\xi}\,d\xi = \frac{\text{constant}}{\mu}\left\{e^{-i\mu a/2} - e^{i\mu a/2}\right\}.$$

The intensity is then

$$|u|^2 = \text{constant}\left(\frac{\sin \mu a/2}{\mu}\right)^2.$$

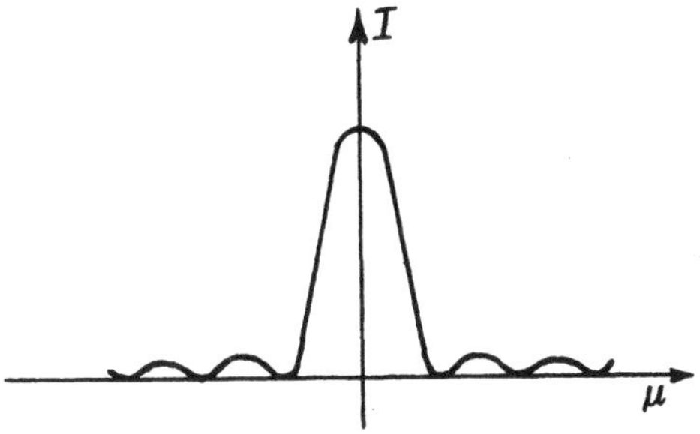

Figure 7.4

If the intensity of the direct ray is called I_0, then

$$|u|^2 = I_0\left(\frac{\sin \mu a/2}{\mu a/2}\right)^2. \qquad [7.9]$$

Thus, we have minima for

$$\frac{a}{2}\mu = \pm\pi,\ \pm 2\pi,\ \ldots,$$

or when

$$a(\sin\psi - \sin\varphi) = n\lambda .$$

2. *Diffraction from a circular aperture.* The circle is placed in the ξ-η plane. Then, Eq. [7.8′] becomes

$$u = \text{constant}\cdot\int_{\text{circle}} e^{-ik[\xi(\alpha_0+\alpha_1)+\eta(\beta_0+\beta_1)]}\,\mathrm{d}\xi\,\mathrm{d}\eta .$$

We now assume that the light is incident perpendicular to the diaphragm so that $\alpha_1=\beta_1=0$ and the z axis becomes the symmetry axis for the entire phenomenon. Furthermore, we introduce the diffraction angle ϑ, and since, on symmetry grounds, we see that u is independent of azimuth, we set $\beta_0 = 0$ so that $\alpha_0 = \sin\vartheta$. There follows

$$u = \text{constant} \cdot \int_{\text{circle}} e^{-ik\xi \sin\vartheta} \mathrm{d}\xi\,\mathrm{d}\eta\ .$$

Here, we must integrate ξ over the range $-\bar{\xi} < \xi < \bar{\xi}$ and η over the range $-a < \eta < a$. If we perform the first integration, there follows

$$u = \text{constant} \cdot \int \frac{1}{\sin\vartheta} \sin(k\bar{\xi}\sin\vartheta)\,\mathrm{d}\eta\ .$$

Now, however, $\bar{\xi}$ can be expressed in terms of the radius of the circle and a polar angle:

$$\bar{\xi} = a\cos\beta\ ,$$

$$\eta = a\sin\beta\ .$$

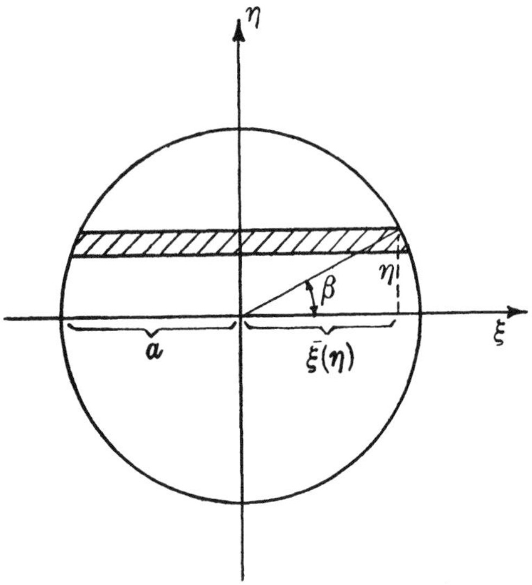

Figure 7.5

Upon substitution there follows

$$u = \frac{\text{constant}}{\sin\vartheta} \int_{-\pi/2}^{+\pi/2} \sin(ka\sin\vartheta\cos\beta)\cos\beta\,\mathrm{d}\beta\ .$$

We know, however, that the Bessel function of order one is defined by

$$J_1(z) = \frac{1}{\pi}\int_{-\pi/2}^{+\pi/2} \sin(z\cos\beta)\cos\beta\, d\beta\,.$$

Thus, for u we get

$$u = \text{constant}\,\frac{J_1(ka\sin\vartheta)}{\sin\vartheta}\,,$$

and for the intensity

$$I \propto |u|^2 = \text{constant}\,\frac{J_1^2(ka\sin\vartheta)}{\sin^2\vartheta}\,.$$

For small z, $J_1(z)/z \sim 1$, so that if the intensity of the undiffracted ray is denoted by I_0, we have the following:

$$I = I_0\left(\frac{J_1(ka\sin\vartheta)}{ka\sin\vartheta}\right)^2. \qquad [7.10]$$

The first zero of $J_1(z)$ is at $z = 1.22\pi$. Consequently, the first dark ring occurs at the diffraction angle ϑ_1, where

$$\sin\vartheta_1 = \frac{1.22\pi}{ka} = 0.61\frac{\lambda}{a}\,. \qquad [7.10']$$

Asymptotically, the Nth zero of $J_1(z)$ is at $z_N \sim \pi(N+\frac{1}{4})$, so that for the minimum it follows that

$$\sin\vartheta_N \sim \left(N+\frac{1}{4}\right)\frac{\lambda}{2a}\,.$$

3. *Diffraction from a grating.* By a grating we mean a system of m (infinitely narrow) equally spaced, adjacent slits. Thus, the integral in Eq. [7.8′] goes over into the following sum:

$$u = \text{constant}\sum_{p=0}^{m-1} e^{-ipdk(\alpha_0+\alpha_1)}\,,$$

or, if we again set

$$\mu = k(\alpha_0 + \alpha_1) = k(\sin\psi - \sin\varphi)$$

(see Fig. 7.3),

$$u = \text{constant} \sum_{p=0}^{m-1} e^{-ip\mu d} = \text{constant} \frac{e^{-im\mu d} - 1}{e^{-i\mu d} - 1}.$$

Thus,

$$I \propto |u|^2 = \text{constant} \frac{1 - \cos m\mu d}{1 - \cos \mu d}$$

$$= \text{constant} \left(\frac{\sin m\mu d/2}{\sin \mu d/2}\right)^2. \qquad [7.11]$$

The principal maxima of order N occur when the numerator and denominator both vanish:

$$\frac{\mu d}{2} = N\pi, \quad (\alpha_1 + \alpha_2)\frac{d}{\lambda} = N. \qquad [7.11']$$

At principal maxima $I \sim m^2$. The intensity vanishes for

$$\frac{m\mu d}{2} = N\pi \quad \text{and} \quad \sin\frac{\mu d}{2} \neq 0.$$

From this, we conclude that the shape of the diffraction pattern is as shown in Fig. 7.6. If the width of the slits

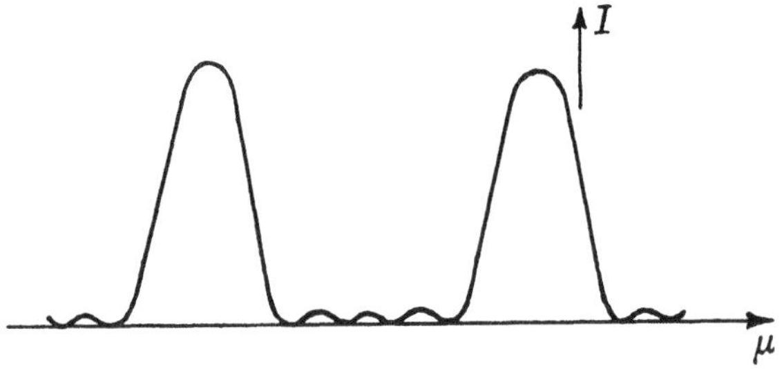

Figure 7.6

in the grating is taken into consideration, then we obtain the product of the formulas for slit and grating. Gratings are the most important instruments for spectral analysis. They permit the measurement of wavelengths.

Technically, two neighboring wavelengths are considered to be "resolved" if the principal maximum of one falls exactly on the first minimum of the other.

According to Eq. [7.11′], $(\sin\varphi - \sin\psi)(d/\lambda) = N$, so that for normal incidence the principal maximum of Nth order occurs at the diffraction angle

$$\sin\psi = \frac{\lambda N}{d},$$

and the first zero after that is separated from it by the angular interval

$$\Delta(\sin\psi) = \frac{\lambda}{dm}.$$

On the other hand, if the wavelength is varied then the center of the principal maximum is displaced by

$$\Delta'(\sin\psi) = \frac{\Delta\lambda}{d} N.$$

If we are to have $\Delta' = \Delta$, then

$$A \equiv \frac{\lambda}{\Delta\lambda} = Nm.$$

Thus, the resolving power $A \equiv \lambda/\Delta\lambda$ is equal to the product of the order times the number of lines in the grating.

4. *The resolving power of a microscope.* The Fraunhofer theory shows how the concept of resolving power can be defined. We consider the example of the microscope.

Luminous object. The point P can no longer be differentiated from Q if P is so close to Q that P' falls within the principal maximum of the diffraction pattern of Q.

If the object plane E is exactly imaged at the image plane E', then the sine condition

$$ln \sin\delta = l'n' \sin\delta' \simeq l'n'\delta'$$

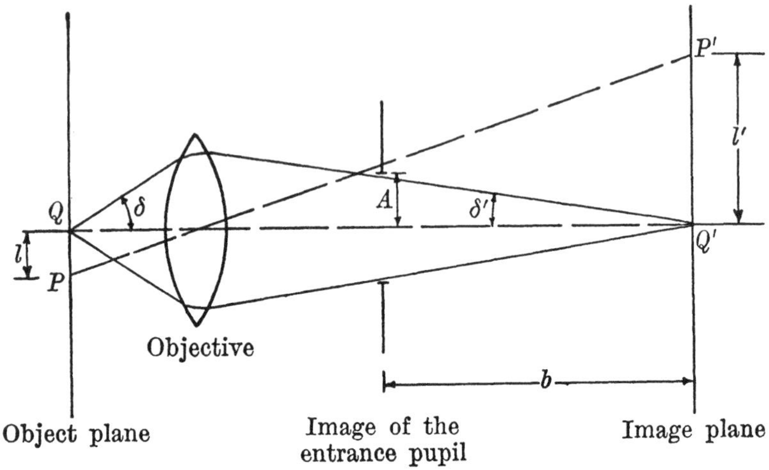

Figure 7.7

must be satisfied. In order that P be differentiable from Q, with the image of the entrance pupil as the diffracting aperture, one must have

$$\frac{l'}{b} > \varphi' = 0.61\,\frac{\lambda'}{A}, \quad \text{but } \frac{A}{b} = \delta',$$

so that

$$l' \geqslant 0.61\,\frac{\lambda_0}{n'\delta'},$$

which, with the above-mentioned sine condition, gives

$$l \simeq \frac{l'n'\delta'}{n \sin\delta} > 0.61\,\frac{\lambda_0}{n \sin\delta}.$$

Illuminated object. A grating with grating constant d is assumed as the "standard" object. Many orders of the diffraction pattern produced by the grating must be included for an image to be obtained. Let φ be the angle of the first principal maximum. Thus, two coherent light sources are formed at C and C_1 in the focal plane, and these sources produce a system of lines with $d' = \lambda b/A$ at

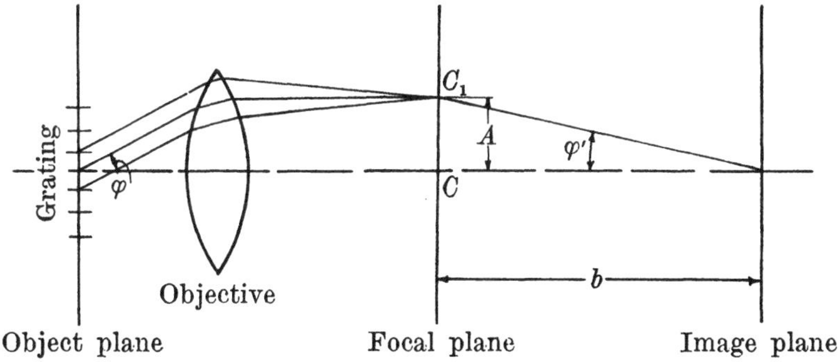

Figure 7.8

the image plane E' (see Fresnel's experiment, Fig. 6.1). Therefore, we get an image whose structure is similar to that of the object only if at least one diffraction maximum of the grating is included. Thus,

$$\varphi < \tfrac{1}{2}\cdot(\text{opening angle}) = \delta\,,$$

$$\sin\varphi = \frac{\lambda_0}{nd} < \sin\delta\,,$$

$$d > \frac{\lambda_0}{n\sin\delta}\,,$$

where $n \sin\delta$ is called the numerical aperture. A factor of $\frac{1}{2}$ is gained when the system is arranged so that the direct ray hits the edge of the lens. If one is satisfied with only the *indication* of a particle without wishing to know its shape, then it is only the light intensity that determines how small the particle may be.

The concept of resolving power can also be applied to other optical instruments. Thus, the calculation for a step-

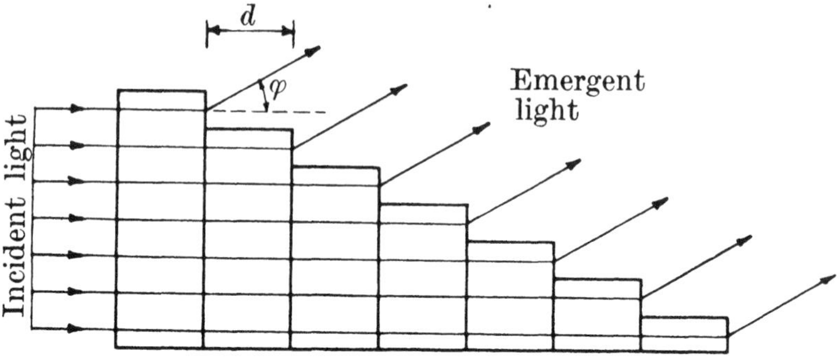

Figure 7.9

grating (Michelson's transmission echelon) yields [A-4]

$$A = md\left(\frac{n-1}{\lambda} - \frac{\mathrm{d}n}{\mathrm{d}\lambda}\right),$$

where m is the number of layers. (For additional examples, see the Supplementary Bibliography.)

c. Fresnel diffraction

We now assume the light source L and the field point P to be finite distances from the diffracting aperture. Then, in Eq. [7.8] terms up to at least the quadratic ones must be considered. At the same time, we locate the origin of the coordinate system at the point of intersection of the connecting line LP with the plane of the diffracting aperture. (For fixed L we have a different coordinate system for each point P.) In this way we achieve the result that in Eq. [7.8] the linear terms in Φ drop out. If we now set $y_1 = y_0 = 0$, then $\alpha = \alpha_0 = \sin\delta$, and there remains

$$\left.\begin{aligned} \Phi &= \frac{1}{2}\left(\frac{1}{R_0} + \frac{1}{R_1}\right)(\xi^2 \cos^2\delta + \eta^2) \\ u(P) &= -\frac{ik\cos\delta}{2\pi R_0 R_1}\, e^{ik(R_0+R_1)} \int e^{ik\Phi}\, \mathrm{d}\xi\, \mathrm{d}\eta \end{aligned}\right\}. \qquad [7.12]$$

We make the following change of variables:

$$\left.\begin{aligned} k\left(\frac{1}{R_1}+\frac{1}{R_0}\right)\xi^2\cos^2\delta &= \pi u^2 \\ k\left(\frac{1}{R_1}+\frac{1}{R_0}\right)\eta^2 &= \pi v^2 \end{aligned}\right\}, \qquad [7.13]$$

so that

$$k\left(\frac{1}{R_0}+\frac{1}{R_1}\right)\cos\delta\,\mathrm{d}\xi\,\mathrm{d}\eta = \pi\,\mathrm{d}u\,\mathrm{d}v\,,$$

with which Eq. [7.12] becomes

$$u(P) = -\frac{i}{2}\frac{e^{ik(R_0+R_1)}}{R_0+R_1}\int e^{i(\pi/2)u^2}\mathrm{d}u\int e^{i(\pi/2)v^2}\mathrm{d}v\,. \qquad [7.14]$$

The limits of integration in Eq. [7.14] are easily transformed. If the original limits were $\pm\infty$ (the case of a slit), then in the transformed integral they are also $\pm\infty$. We have

$$\int_{-\infty}^{+\infty} e^{i(\pi/2)v^2}\mathrm{d}v = 2\int_0^{\infty} e^{i(\pi/2)u^2}\mathrm{d}u = \frac{2}{\sqrt{2}}e^{i(\pi/4)} = 1+i\,.$$

The same integral between finite limits,

$$\int_{s-D}^{s} e^{i(\pi/2)u^2}\mathrm{d}u\,, \qquad [7.15]$$

is called Fresnel's integral and is tabulated.

Example. Diffraction from the edge of a semi-infinite plane.

The second integral in Eq. [7.14] clearly has the limits $\pm\infty$, while the first integral is to be taken from a finite limit to infinity. For the discussion of the integral [7.15], we consider the curve defined as follows:

$$X = \int_0^s \cos\left(\frac{\pi}{2}u^2\right)\mathrm{d}u\,, \qquad Y = \int^s \sin\left(\frac{\pi}{2}u^2\right)\mathrm{d}u\,. \qquad [7.16]$$

Since $(\mathrm{d}X/\mathrm{d}s)^2 + (\mathrm{d}Y/\mathrm{d}s)^2 = 1$, then s is the arc length. If we set $\mathrm{d}X/\mathrm{d}s = \cos\tau$, $\mathrm{d}Y/\mathrm{d}s = \sin\tau$, then τ is the angle the tangent makes with the X axis. Then, however, the radius of curvature is $\varrho = \mathrm{d}s/\mathrm{d}\tau$, and it follows from Eq. [7.16] that

$$\varrho = \frac{1}{\pi s}\,. \qquad [7.17]$$

The curve which is defined by Eq. [7.16] is the familiar Cornu spiral. It can be constructed approximately, for example, from Eq. [7.17]. (See Fig. 7.10.)

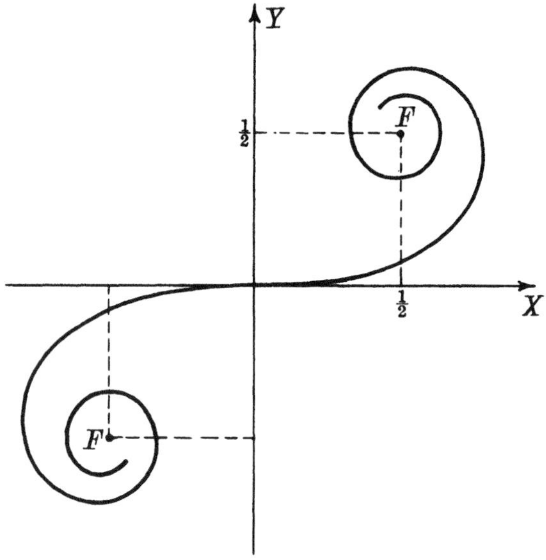

Figure 7.10

If we consider the integral of Eq. [7.15], then we see that

$$\left.\begin{aligned} \left| \int_0^s e^{i(\pi/2)u^2}\,\mathrm{d}u \right|^2 &= X^2 + Y^2 \\ \left| \int_{s'}^s e^{i(\pi/2)u^2}\,\mathrm{d}u \right|^2 &= (X - X')^2 + (Y - Y')^2 \end{aligned}\right\}. \qquad [7.18]$$

In Fig. 7.11, the coordinate system has been chosen so that the projection of LP on the plane falls along the x axis, which is perpendicular to the edge of the plane. We thus

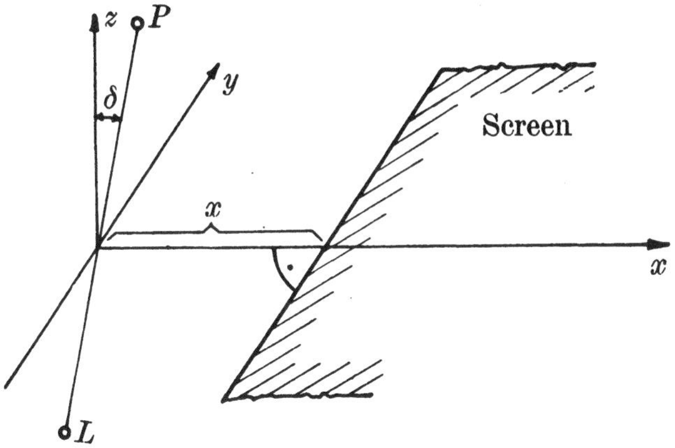

Figure 7.11

see that the limits for ξ are $-\infty$ and x, so that from [7.13] the limits for u are $-\infty$ and $x \cos \delta \{k(R_0+R_1)/\pi R_0 R_1\}^{\frac{1}{2}}$. From Eq. [7.18], we recognize that the intensity ratio $2I/I_0$ [where I_0 corresponds to undisturbed light propagation, $e^{ik(R_0+R_1)}/(R_0+R_1)$] for $u(P) = u(s)$ is obtained by traveling along the Cornu spiral a distance s from the origin (and with the proper sign), and then taking the square of the distance from the point thus obtained to the end of the spiral at F. The case $s<0$ corresponds to geometrical shadow, while $s>0$ corresponds to the geometrically illuminated side. In particular, for

$$s = +\infty, \qquad I/I_0 = 1\,;$$

$$s = 0, \qquad I/I_0 = \tfrac{1}{4} \qquad \text{(shadow boundary)}\,;$$

$$s = -\infty, \qquad I/I_0 = 0\,.$$

The result is shown in Fig. 7.12. On the shadow side, the

intensity decreases monotonically; on the light side, however, the intensity oscillates.

If instead of a semi-infinite plane we have a slit of width d,

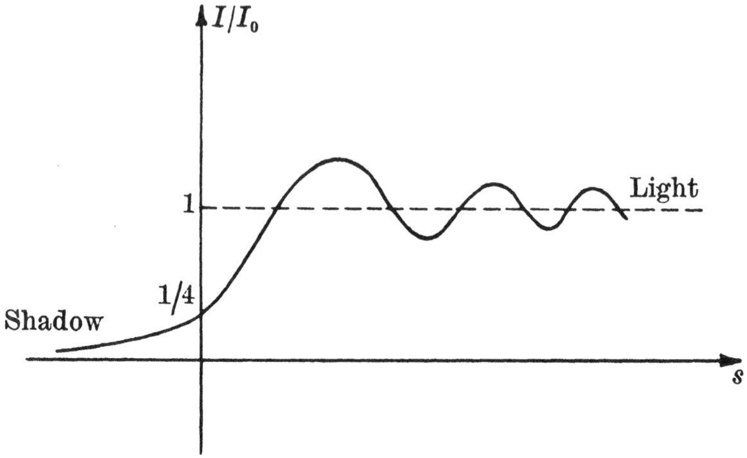

Figure 7.12

then everything is calculated as before except that the limits for the second integration are now $s-D$ and s, where

$$s = x \cos\delta\{k(R_0+R_1)/\pi R_0 R_1\}^{\frac{1}{2}}$$

as before, and

$$D = d \cos\delta\{k(R_0+R_1)/\pi R_0 R_1\}^{\frac{1}{2}}.$$

Chapter 3. Maxwell's Theory

8. FOUNDATIONS OF THE THEORY

It turns out that the scalar wave optics developed in the previous chapter does not agree with certain experimental facts. That is, certain optical phenomena can be demonstrated which can be explained only on the basis of the concept of transverse vector waves (for example, birefraction and polarization by reflection). A light beam is said to be polarized if the direction of the vector which will now be associated with the propagation of light can be specified. Experimentally, it is found that:

1. Interference between light beams of like polarization is the same as for unpolarized light.
2. Two light beams with polarizations perpendicular to one another do not interfere. (From this it follows that light has no longitudinal component.) Furthermore, there is no interference if the two originally perpendicularly polarized components of a light beam are brought into coincidence by repolarization.
3. If uniformly polarized light is repolarized several times, then there is interference only if finally the light is again uniformly polarized.

The (experimentally determined) pure transversality of light waves and their (experimentally determined) velocity of propagation suggest the conjecture that the hypothesized oscillating light vector is to be identified with the magnetic or electric field vector, $\boldsymbol{H}$ or $\boldsymbol{E}$. As shown in *Electro-*

dynamics,[1] these fields also satisfy a wave equation. For a charge-free region, and for processes varying in time, the Maxwell equations (for $\mu = 1$) hold:

$$\left.\begin{array}{ll} \dfrac{\varepsilon}{c}\dfrac{\partial \boldsymbol{E}}{\partial t} = \operatorname{curl} \boldsymbol{H}, & \operatorname{div} \boldsymbol{E} = 0 \\ \dfrac{1}{c}\dfrac{\partial \boldsymbol{H}}{\partial t} = -\operatorname{curl} \boldsymbol{E}, & \operatorname{div} \boldsymbol{H} = 0 \end{array}\right\} . \qquad [8.1]$$

From the familiar vector identity

$$\nabla^2 \boldsymbol{A} = \operatorname{grad} \operatorname{div} \boldsymbol{A} - \operatorname{curl} \operatorname{curl} \boldsymbol{A} ,$$

there follows

$$\left.\begin{array}{l} \nabla^2 \boldsymbol{E} - \dfrac{\varepsilon}{c^2}\dfrac{\partial^2 \boldsymbol{E}}{\partial t^2} = 0 \\ \nabla^2 \boldsymbol{H} - \dfrac{\varepsilon}{c^2}\dfrac{\partial^2 \boldsymbol{H}}{\partial t^2} = 0 \end{array}\right\} , \qquad [8.2]$$

from which the phase velocity is seen to be

$$v = \frac{c}{\sqrt{\varepsilon}} . \qquad [8.3]$$

The successful results which follow from the identification of the "light vector" with $\boldsymbol{E}$ justify this hypothesis. It turns out that what was formerly denoted as the "light vector" corresponds to the quantity $\boldsymbol{H}$. For this choice, the "plane of polarization" is to be understood as the plane which contains $\boldsymbol{H}$ and the wave normal [A-2]. However, most of the effects of light are to be attributed to the effects of $\boldsymbol{E}$ (see p. 83) since, in accordance with the general electromagnetic law of force, $\boldsymbol{E}$ exerts a force on stationary as well as on moving charges, while $\boldsymbol{H}$ acts only on moving charges. Thus, we will always operate with $\boldsymbol{E}$. In order to preserve agreement with the previous results, we must set

$$n = \sqrt{\varepsilon}$$

[1] W. Pauli, *Lectures in Physics: Electrodynamics* (M.I.T. Press, Cambridge, Mass., 1972).

in accordance with Eq. [8.3]. In an electromagnetic wave, $\boldsymbol{E}$ and $\boldsymbol{H}$ are perpendicular to one another and to the direction of the ray.

We now consider a plane wave which satisfies Eq. [8.2] and which propagates in the x direction. Thus, all field components depend only on the variable $(x-vt)$, so that

$$\partial/\partial y = \partial/\partial z = 0\,, \qquad \partial/\partial t = -c\varepsilon^{-\frac{1}{2}}\partial/\partial x\,.$$

From this, it is immediately seen that with this hypothesis two separate solutions of Eqs. [8.1] and [8.2] are obtained:

$$\left.\begin{array}{l} (1)\ E_z\,,\quad H_y \neq 0\,, \\ \qquad \text{the remaining field components being zero} \\ (2)\ E_y\,,\quad H_z \neq 0\,, \\ \qquad \text{the remaining field components being zero} \end{array}\right\}. \qquad [8.4]$$

In these cases, the Maxwell equations [8.1] give

$$\left.\begin{array}{ll} (1) & H_y = -\sqrt{\varepsilon}\,E_z \\ (2) & H_z = +\sqrt{\varepsilon}\,E_y \end{array}\right\}. \qquad [8.5]$$

These two solutions correspond to the two types of polarization for plane waves; the solutions can be linearly superimposed, and the form thereby obtained is still a solution of Maxwell's equations.

As is well known, the law of energy conservation also follows from Maxwell's equations:

$$\operatorname{div}\boldsymbol{S} + \frac{\partial W}{\partial t} = 0\,, \qquad [8.6]$$

where $\boldsymbol{S} = (c/4\pi)(\boldsymbol{E}\times\boldsymbol{H})$, $W = 1/8\pi(\varepsilon\boldsymbol{E}^2 + \mu\boldsymbol{H}^2)$.

Maxwell's equations [8.1] can also be written in integral form. In that case, the formulas in Eqs. [8.1] which contain the curl are equivalent to

$$\left.\begin{aligned}\oint H_s \mathrm{d}s &= \frac{1}{c}\int \varepsilon \dot{E}_n \mathrm{d}f \\ \oint E_s \mathrm{d}s &= -\frac{1}{c}\int \dot{H}_n \mathrm{d}f\end{aligned}\right\} . \quad [8.7]$$

Equations [8.7] are used to advantage when ε becomes discontinuous (as, for example, at a surface), in which case Eqs. [8.1] break down. In such a case, shown in Fig. 8.1,

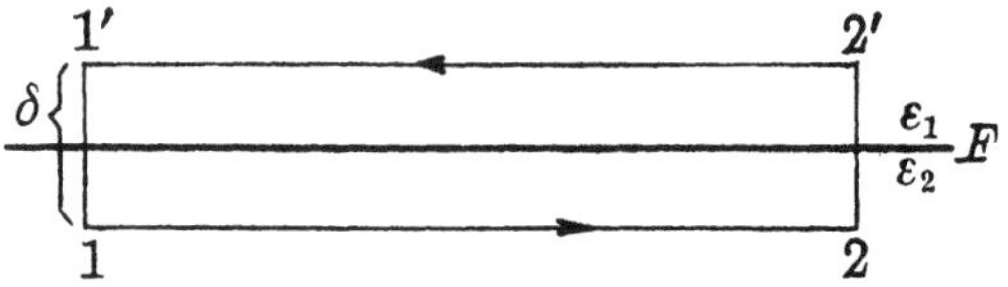

Figure 8.1

ε has a finite discontinuity at the surface F, and the parts of the integrals over the lengths δ (δ small) are of the same order of magnitude as δ. However, the term on the right in the first of Eqs. [8.7] is also of order of magnitude δ, since $\dot{E}_n$ certainly remains finite. Thus, there follows

$$\int_1^2 E_s \mathrm{d}s - \int_{1'}^{2'} E_s \mathrm{d}s = O(\delta),$$

and from the limit $\delta \to 0$ we conclude that the parallel component of $\boldsymbol{E}$ (and $\boldsymbol{H}$) remains continuous across a surface at which ε changes discontinuously:

$\boldsymbol{E}_{\parallel}$, $\boldsymbol{H}_{\parallel}$ *are continuous across a surface where* ε *changes discontinuously* . [8.8]

Up to this point, we have proceeded as if ε were a material constant, although this is not the case. This can be taken into consideration if the time dependence is assumed to be periodic, $\sim e^{-i\omega t}$. Then, $\partial/\partial t \sim -i\omega$, and there follows from Eqs. [8.1] that

$$\left.\begin{aligned} -\frac{\varepsilon i\omega}{c}\boldsymbol{E} &= \operatorname{curl}\boldsymbol{H} \\ \frac{i\omega}{c}\boldsymbol{H} &= \operatorname{curl}\boldsymbol{E} \end{aligned}\right| . \qquad [8.9]$$

This can next be transformed into the form of the time-independent wave equation (in analogy to the equation $\nabla^2 u + k^2 u = 0$ of Chapter 2):

$$\left.\begin{aligned} \nabla^2\boldsymbol{E} + \frac{\varepsilon\omega^2}{c^2}\boldsymbol{E} &= 0 \\ \nabla^2\boldsymbol{H} + \frac{\varepsilon\omega^2}{c^2}\boldsymbol{H} &= 0 \end{aligned}\right| . \qquad [8.10]$$

Whether the system of equations [8.9] or [8.1] is more appropriate for a specific case must be decided on the basis of the experimental setup.

9. NONABSORBING MEDIA (FRESNEL'S FORMULAE)

The system of equations [8.10] possesses solutions of the form

$$\boldsymbol{E} = \boldsymbol{E}_0 e^{i(\boldsymbol{k}\cdot\boldsymbol{x} - \omega t)} \qquad [9.1]$$

(and analogously for $\boldsymbol{H}$), if we set

$$k^2 = \frac{\varepsilon\omega^2}{c^2}.$$

The phase velocity is then $v = c/\sqrt{\varepsilon}$, and, as was noted previously, $n = \sqrt{\varepsilon}$ (isotropic medium).

We consider the case where ε changes discontinuously across a plane. Since Eqs. [8.10] possess solutions of the form [9.1], we choose the initial conditions for the general optical wave field in such a manner that they correspond to an incident plane wave in the x-z plane, making an angle α with the z axis. (See Fig. 9.1.) From Eq. [8.4] we have seen that the system of Maxwell's equations—in the case that plane waves may be assumed as solutions [2]—

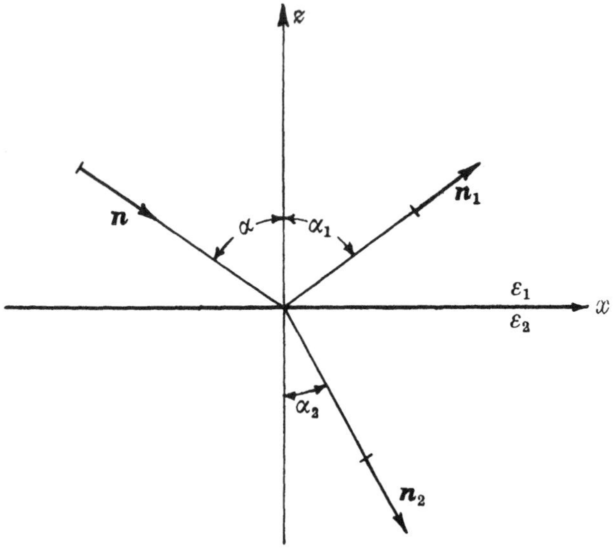

Figure 9.1

possesses two completely different solutions. Thus, here too it will be necessary to treat these two cases separately:

Case 1. ***E*** perpendicular to the plane of incidence,

Case 2. ***E*** parallel to the plane of incidence.

First, however, we can establish that the continuity of the phase at the bounding surface, as shown in Section 6, results in the fact that ω in one medium is the same as ω in the other, and furthermore that $\alpha = \alpha_1$ and $\sqrt{\varepsilon_1} \sin \alpha_1 = \sqrt{\varepsilon_2} \sin \alpha_2$, or $\sin \alpha = n \sin \alpha_2$, where $n = \sqrt{\varepsilon_1/\varepsilon_2}$. Thus, the unit vectors $\boldsymbol{n}$, $\boldsymbol{n}_1$, and $\boldsymbol{n}_2$ of the incident, reflected, and

[2] This is to be expected because of the chosen initial condition.

refracted waves, respectively, become

$$\boldsymbol{n} = \{\sin\alpha, 0, -\cos\alpha\}, \quad \boldsymbol{n}_1 = \{\sin\alpha, 0, \cos\alpha\},$$
$$\boldsymbol{n}_2 = \{\sin\alpha_2, 0, -\cos\alpha_2\}.$$

Case 1. ***E perpendicular to the plane of incidence***

Form of solution: All fields are multiplied by $e^{-i\omega t}$. Furthermore, $\boldsymbol{k}_i = k\boldsymbol{n}_i$ and $\boldsymbol{n}_i \cdot \boldsymbol{x} = s_i$. Then, clearly,

$$\left.\begin{aligned} E_y &= Ae^{ik_1 s} \\ H_x &= \sqrt{\varepsilon_1}\, A\cos\alpha\, e^{ik_1 s} \\ H_z &= \sqrt{\varepsilon_1}\, A\sin\alpha\, e^{ik_1 s} \end{aligned}\right\}, \quad \text{incident wave;}$$

$$\left.\begin{aligned} E_y &= Be^{ik_1 s_1} \\ H_x &= -\sqrt{\varepsilon_1}\, B\cos\alpha\, e^{ik_1 s_1} \\ H_z &= \sqrt{\varepsilon_1}\, B\sin\alpha\, e^{ik_1 s_1} \end{aligned}\right\}, \quad \text{reflected wave;}$$

$$\left.\begin{aligned} E_y &= Ce^{ik_2 s_2} \\ H_x &= \sqrt{\varepsilon_2}\, C\cos\alpha_2\, e^{ik_2 s_2} \\ H_z &= \sqrt{\varepsilon_2}\, C\sin\alpha_2\, e^{ik_2 s_2} \end{aligned}\right\}, \quad \text{refracted wave.}$$

For the relation between $|\boldsymbol{E}|$ and $|\boldsymbol{H}|$, see Eq. [8.5]. On the basis of the principle of [8.8], since $k_1 s = k_1 s_1 = k_2 s_2$ for $z = 0$, there follows

from the continuity of E_y: $A + B = C$,

from the continuity of H_x: $(A - B)\cos\alpha = Cn\cos\alpha_2$,

so that

$$\left.\begin{aligned} \frac{B}{A} &= -\frac{\sin(\alpha-\alpha_2)}{\sin(\alpha+\alpha_2)} \\ \frac{C}{A} &= \frac{2\cos\alpha\sin\alpha_2}{\sin(\alpha+\alpha_2)} \end{aligned}\right\}. \qquad [9.2]$$

Case 2. ***E parallel to the plane of incidence***

We take $\boldsymbol{H}$ in the $-y$ direction, so that $\boldsymbol{E}$ now plays the role of $\boldsymbol{H}$ in Case 1 above.

Form of solution: (For the relation between $|\boldsymbol{E}|$ and $|\boldsymbol{H}|$, see Eq. [8.5].)

$$\left.\begin{aligned} H_y &= -Ae^{ik_1 s} \\ E_x &= \varepsilon_1^{-1/2} A \cos\alpha\, e^{ik_1 s} \\ E_z &= \varepsilon_1^{-1/2} A \sin\alpha\, e^{ik_1 s} \end{aligned}\right\}, \quad \text{incident wave;}$$

$$\left.\begin{aligned} H_y &= -Be^{ik_1 s_1} \\ E_x &= -\varepsilon_1^{-1/2} B \cos\alpha\, e^{ik_1 s_1} \\ E_z &= \varepsilon_1^{-1/2} B \sin\alpha\, e^{ik_1 s_1} \end{aligned}\right\}, \quad \text{reflected wave;}$$

$$\left.\begin{aligned} H_y &= -Ce^{ik_2 s_2} \\ E_x &= \varepsilon_2^{-1/2} C \cos\alpha_2\, e^{ik_2 s_2} \\ E_z &= \varepsilon_2^{-1/2} C \sin\alpha_2\, e^{ik_2 s_2} \end{aligned}\right\}. \quad \text{refracted wave.}$$

The continuity of the parallel components yields

$$A + B = C,$$

$$(A - B)\cos\alpha = \frac{C}{n}\cos\alpha_2,$$

from which

$$\left.\begin{aligned} \frac{B}{A} &= \frac{\tan(\alpha - \alpha_2)}{\tan(\alpha + \alpha_2)} \\ \frac{C}{A} &= \frac{2\sin 2\alpha}{\sin 2\alpha + \sin 2\alpha_2} \end{aligned}\right\}. \qquad [9.3]$$

Discussion. From *Electrodynamics*,[3] the intensity I of an electromagnetic wave is proportional to the Poynting vector $\boldsymbol{S}$:

$$\boldsymbol{S} = \frac{c}{4\pi}\boldsymbol{E}\times\boldsymbol{H}, \qquad I = |\boldsymbol{S}|.$$

It is also known that for a plane wave (which is the case

[3] W. Pauli, *Electrodynamics.*

here), $\boldsymbol{E}$ is normal to $\boldsymbol{H}$. Thus, from Eq. [8.5], it follows that for our case

$$I \propto |\boldsymbol{E}||\boldsymbol{H}| = |\boldsymbol{H}|^2/n = n|\boldsymbol{E}|^2. \qquad [9.4]$$

Thus, as long as total reflection does not occur, so that the exponential factor $e^{i(kz-\omega t)}$ has magnitude unity, the ratios of the reflected and refracted intensities, respectively, to that of the incident intensity for the first case are

$$I_{\text{refl}}:I_{\text{inc}} = |B/A|^2, \quad I_{\text{refr}}:I_{\text{inc}} = n|C/A|^2 \qquad [9.5]$$

from Eq. [9.2], while for the second case

$$I_{\text{refl}}:I_{\text{inc}} = |B/A|^2, \quad I_{\text{refr}}:I_{\text{inc}} = |C/A|^2/n \qquad [9.6]$$

from Eq. [9.3].

Of course, at normal incidence both formulas become identical:

$$I_{\text{refl}}:I_{\text{inc}} = \{(n-1)/(n+1)\}^2, \quad I_{\text{refr}}:I_{\text{inc}} = 4n/(n+1)^2. \qquad [9.7]$$

At glancing incidence, $\alpha = \pi/2$, it follows in both cases that $I_{\text{refl}}:I_{\text{inc}} = 1$. Furthermore, for the second case, it is seen from Eq. [9.3] that there exists an angle α_ϱ such that $I_{\text{refl}}:I_{\text{inc}}$ is zero. This is the case when

$$\alpha_\varrho + \alpha_2 = \pi/2 \qquad [9.8]$$

(since $\tan \pi/2 = \infty$); this is the case when the refracted ray and the (nonexistent) reflected ray would be perpendicular to one another. This is *Brewster's law*. Thus, only one component of unpolarized light incident on a reflecting surface at an angle α_ϱ, as determined by Eq. [9.8], is reflected. Consequently, the reflected light is linearly polarized.

Since the component of the vector $\boldsymbol{E}$ which oscillates normal to the plane of incidence experiences different effects than the parallel component, then oblique linearly polarized light remains linearly polarized after refraction and reflection, although in general its plane of polarization is rotated.

Somewhat different relations are obtained for the case of *total reflection.* Since formulas [9.2] and [9.3] are purely formal solutions of Maxwell's equations, then they of course remain valid for this case too. On the contrary, formulas [9.5] and [9.6] for the intensities are no longer valid, since the exponential factors are no longer necessarily of magnitude one.

In order that total reflection occur, in accordance with Chapter 1 we must assume that $n<1$ and that α is sufficiently large so that

$$\sin\alpha = n \sin\alpha_2$$

no longer has a real solution for α_2. Thus,

$$\alpha > \alpha_l\,, \quad \text{where} \quad \sin\alpha_l = n\,. \qquad [9.9]$$

The angle α_l defined in this way is called the *limiting angle for total reflection.*

As can easily be shown, for $\sin\alpha_2$ to be real and to be greater than unity, α_2 must necessarily be of the form

$$\alpha_2 = \pi/2 - i\beta\,, \qquad [9.10]$$

where β is real (the sign in front of β is negative by convention only). With this substitution, if we completely write out the expressions for the field components (including the exponential factors), we obtain

$$\left.\begin{array}{r} E_v \\ -H_v \end{array}\right\} = B e^{-i(\omega t - k_1 s_1)} = B e^{-i(\omega t - k_1 x \sin\alpha - k_1 z \cos\alpha)} \quad \text{reflected wave}\,, \qquad [9.11]$$

and

$$\left.\begin{array}{r} E_v \\ -H_v \end{array}\right\} = C e^{-i(\omega t - k_2 s_2)} = C e^{-i\omega t + ik_2 x \cosh\beta}\, e^{k_2 z \sinh\beta} \quad \text{refracted wave}\,. \qquad [9.12]$$

This represents a so-called surface wave, since there is periodicity only in the x direction, while in the z direction there is an exponential falloff. Upon computation, the Poynting vector is seen to be parallel to the plane of refraction. For both cases of polarization there follows $I_{\text{refl}}:I_{\text{inc}}=1$, so that all the light is reflected. On the contrary, it is noticed that for reflection the field components for the two cases have *different* complex factors; that is, linearly polarized light oblique to the plane of incidence becomes elliptically polarized after reflection. The complex factors are just those values for B/A given by Eqs. [9.2] and [9.3] when the substitution [9.10] is used.

With Fresnel's formulae, a modification of the geometric ray theory has been obtained, such that even for the case of total reflection a wave in the second medium must be assumed, in accordance with Eq. [9.12]. It can be shown experimentally that (as is demanded by the theory) light is always transmitted through an interrupting layer of finite thickness with $n<1$, even if total reflection should occur according to the geometric theory.

10. ABSORBING MEDIA (OPTICS OF METALS)

We start once again from the Maxwell equations, except that we assume that the medium possesses an electrical conductivity. Then, for processes which are periodic with $e^{-i\omega t}$, the Maxwell equations become

$$\left.\begin{aligned} \operatorname{curl}\boldsymbol{H} &= \left(-\frac{\varepsilon}{c}i\omega+\frac{4\pi\sigma}{c}\right)\boldsymbol{E} \\ \operatorname{curl}\boldsymbol{E} &= \frac{1}{c}\,i\omega\boldsymbol{H} \end{aligned}\right\} \qquad [10.1]$$

(since $\partial/\partial t\sim-i\omega$). We can rewrite the first of these equations:

$$\operatorname{curl}\boldsymbol{H}=\left(-\frac{\varepsilon}{c}\,i\omega+\frac{4\pi\sigma}{c}\right)\boldsymbol{E}=-\frac{i\omega}{c}\,n'^2\boldsymbol{E}\,,$$

where we have defined

$$n'^2 = \varepsilon' = \varepsilon + i\,\frac{4\pi\sigma}{\omega}\,. \qquad [10.2]$$

Furthermore, if we substitute

$$n' = n(1 + i\varkappa) \qquad (n,\ \varkappa,\ \text{real})\,, \qquad [10.3]$$

then upon comparison with Eq. [10.2] there follows

$$\left.\begin{aligned} n^2(1-\varkappa^2) &= \varepsilon \\ n^2\varkappa &= 2\pi\sigma/\omega \end{aligned}\right\}. \qquad [10.4]$$

Thus, Maxwell's equations of [10.1] become

$$\left.\begin{aligned} \operatorname{curl}\boldsymbol{H} &= -\frac{i\omega}{c}\,\varepsilon'\boldsymbol{E} \\ \operatorname{curl}\boldsymbol{E} &= \quad \frac{i\omega}{c}\,\boldsymbol{H} \end{aligned}\right\}. \qquad [10.5]$$

Equations [10.5] are exactly the same as Eqs. [8.1] written for periodic processes, except that ε' appears everywhere in place of ε. Thus, all the previously developed solutions of the Maxwell equations are also valid here if only we substitute ε' (complex) for ε and n' for n.

For example, in Eq. [9.1] we had found a solution of Maxwell's equations which, when carried over to the present case, becomes

$$\boldsymbol{E} = \boldsymbol{E}_0 e^{i(\boldsymbol{k}\cdot\boldsymbol{x}-\omega t)}\,, \qquad \text{with} \qquad \boldsymbol{x} = \{x,\ 0,\ 0\}\,, \qquad [10.6]$$

where (in this case)

$$k^2 = \frac{\varepsilon'\omega^2}{c^2}\,, \qquad k = \frac{\omega}{c}\,n' = \frac{\omega}{c}\,(1+i\varkappa)\,n\,.$$

Thus, from Eq. [10.6] we get (as a special case of Eq. [8.4])

$$E_z = E_0 e^{-(\varkappa n\omega/c)x} e^{i((n\omega/c)x-\omega t)}\,. \qquad [10.7]$$

From Eq. [8.5], the corresponding magnetic field intensity is

$$H_y = H_0 e^{-(\varkappa n\omega/c)x} e^{i((n\omega/c)x-\omega t)},$$

where

$$H_0 = -n'E_0 . \qquad [10.8]$$

Equation [10.7] represents a damped plane wave whose damping constant is proportional to $\varkappa$. Here too, n (without the prime) can be interpreted as the refractive index, since c/n is the phase velocity. It should be noted that, according to Eq. [10.8], $\boldsymbol{E}$ and $\boldsymbol{H}$ no longer have the same phase, since n' is complex.

If in the above equations we again set $\sigma = 0$, then it follows that $\varkappa = 0$ and that $n = n'$ is real. For the case of metals, however, $\sigma \neq 0$. Indeed, σ is so large that $n^2\varkappa \gg \varepsilon$; that is, $\varkappa \sim 1$ and n is large.

We now consider the case of a linearly polarized plane wave incident on a plane metal surface, with the angle of incidence being α and the angle of refraction α_2. (See Fig. 9.1.) The Fresnel formulae [9.1] and [9.2] for both cases of polarization can, in accordance with the above remarks, be taken over if ε is replaced by ε' and n by n'. Because of the assumptions for metals, it follows that $\cos\alpha_2 \sim 1$.

Case 1. ***E*** *perpendicular to the plane of incidence*

$$\frac{C}{A} = \frac{2\cos\alpha\sin\alpha_2}{\sin(\alpha+\alpha_2)} \simeq \frac{2\cos\alpha}{n'} . \qquad [10.9]$$

Thus, for the refracted wave,

$$\left.\begin{aligned} E_y &= \frac{2A\cos\alpha}{n'} \exp\{i[k_1(x\sin\alpha - nz) - \omega t] + nk_1\varkappa z\} \\ H_x &= \sqrt{\varepsilon_1}\, 2A\cos\alpha \exp\{i[k_1(x\sin\alpha - nz) - \omega t] \\ &\qquad\qquad + n\varkappa k_1 z\} \\ H_z &= \sqrt{\varepsilon_1}\frac{2A}{n'}\cos\alpha\sin\alpha \exp\{i[k_1(x\sin\alpha - nz) - \omega t] \\ &\qquad\qquad + n\varkappa k_1 z\} \end{aligned}\right\} . \qquad [10.10]$$

For $n' \to \infty$, only H_x remains finite (even for $z = 0$).

Case 2. ***E*** *parallel to the plane of incidence*

$$\frac{C}{A} = \frac{2 \sin 2\alpha}{\sin 2\alpha + \sin 2\alpha_2} \simeq 2 . \qquad [10.11]$$

Thus,

$$\left.\begin{aligned} H_y &= -2A \exp\{i[k_1(x \sin\alpha - nz) - \omega t] + n\varkappa k_1 z\} \\ E_x &= \varepsilon_1^{-\frac{1}{2}} \frac{2A}{n'} \exp\{i[k_1(x \sin\alpha - nz) - \omega t] + n\varkappa k_1 z\} \\ E_z &= \varepsilon_1^{-\frac{1}{2}} \sin\alpha \frac{2A}{n'^2} \exp\{i[k_1(x \sin\alpha - nz) - \omega t] + n\varkappa k_1 z\} \end{aligned}\right\} \qquad [10.12]$$

for the refracted wave. Consequently, in the limiting case of a perfect metal ($n' \to \infty$) the electric field intensities go to zero (for $z = 0$ as well). However, it is clear that this must be so, for the current is proportional to ***E***, and we have assumed that the conductivity σ is infinite.

Normal incidence. From Eq. [9.7], there follows

$$\frac{I_{\text{refl}}}{I_{\text{inc}}} = \left|\frac{B}{A}\right|^2 = \left|\frac{n'-1}{n'+1}\right|^2 = \frac{(n-1)^2 + n^2\varkappa^2}{(n+1)^2 + n^2\varkappa^2} .$$

From this follows the important formula for metals

$$1 - \frac{I_{\text{refl}}}{I_{\text{inc}}} = \frac{4n}{(n+1)^2 + n^2\varkappa^2} \simeq \frac{4n}{2n^2} = \frac{2}{n} \sim 2\sqrt{\frac{\omega}{2\pi\sigma}} .$$

In accordance with Eq. [10.4], for $\varkappa \to 1$ there follows

$$1 - \frac{I_{\text{refl}}}{I_{\text{inc}}} = 2\sqrt{\frac{c}{\lambda\sigma}} . \qquad [10.13]$$

Ellipticity of the reflected light. We assume that linearly polarized light, with field amplitudes $A_\perp$ and $A_\parallel$, is incident on the metal surface. Then, from Eqs. [9.2] and [9.3] there follows

$$\left(\frac{B}{A}\right)_\perp = -\frac{\sin(\alpha - \alpha_2)}{\sin(\alpha + \alpha_2)} , \qquad \left(\frac{B}{A}\right)_\parallel = \frac{\tan(\alpha - \alpha_2)}{\tan(\alpha + \alpha_2)} .$$

Thus,

$$\frac{B_\perp}{B_\parallel} = -\frac{A_\perp}{A_\parallel} \frac{\cos(\alpha - \alpha_2)}{\cos(\alpha + \alpha_2)} . \qquad [10.14]$$

Since the quotient of the cosines in Eq. [10.14] is, in general, not real, the reflected light is elliptically polarized. We consider the special case where $A_\perp/A_\parallel = 1$; that is, the plane of incidence and the plane of polarization make an angle of 45°. Furthermore, we set

$$\frac{B_\perp}{B_\parallel} = \varrho e^{i\Delta} . \qquad [10.15]$$

We compute

$$\frac{1+\varrho e^{i\Delta}}{1-\varrho e^{i\Delta}} = -\frac{\sin\alpha \sin\alpha_2}{\cos\alpha \cos\alpha_2} .$$

However,

$$\sin\alpha_2 = \frac{1}{n'} \sin\alpha , \qquad \cos\alpha_2 = \frac{\sqrt{n'^2 - \sin^2\alpha}}{n'} , \qquad [10.16]$$

which, upon substitution, gives

$$\frac{1+\varrho e^{i\Delta}}{1-\varrho e^{i\Delta}} = -\frac{\sin^2\alpha}{\cos\alpha} (n'^2 - \sin^2\alpha)^{-\frac{1}{2}} . \qquad [10.17]$$

From this we see that for

$$\alpha = 0: \qquad \varrho e^{i\Delta} = -1, \qquad \Delta = \pi, \qquad \varrho = 1 ,$$

$$\alpha = \pi/2: \qquad \varrho e^{i\Delta} = +1, \qquad \Delta = 0, \qquad \varrho = 1 ;$$

there is no ellipticity.

The angle of incidence α_{pr} for which $\Delta = \pi/2$ so that $e^{i\Delta} = i$ is called the *principal angle of incidence.* For this case, if we neglect $\sin^2\alpha_{pr}$ compared with n'^2, there follows from Eq. [10.17]

$$\frac{1+i\varrho_{pr}}{1-i\varrho_{pr}} = -\frac{\sin^2\alpha_{pr}}{\cos\alpha_{pr}} \frac{1}{n'} . \qquad [10.18]$$

The quantities ϱ_{pr} and α_{pr} are measured experimentally, so that the optical constants follow from Eq. [10.18]:

$$n' = n(1 + i\varkappa) = -\frac{\sin^2\alpha_{\rm pr}}{\cos\alpha_{\rm pr}}\left(\frac{1 - i\varrho_{\rm pr}}{1 + i\varrho_{\rm pr}}\right),$$

$$\left.\begin{aligned} n &= -\frac{\sin^2\alpha_{\rm pr}}{\cos\alpha_{\rm pr}}\left(\frac{1-\varrho_{\rm pr}^2}{1+\varrho_{\rm pr}^2}\right) \\ \varkappa &= -\frac{2\varrho_{\rm pr}}{1-\varrho_{\rm pr}^2} = \tan 2\psi, \qquad \varrho_{\rm pr} = -\tan\psi \end{aligned}\right\}. \qquad [10.19]$$

It turns out that the "constants" are frequency dependent. They all approach their static values if ω becomes sufficiently small.

11. STANDING WAVES

We consider a plane wave incident on a perfect mirror ($E_\parallel$ vanishes for $z = 0$). We obtain the following for the two cases of polarization:

a. **E** *perpendicular to the plane of incidence:* $B/A = -1$

The superposition of the incident and reflected light waves yields

$$\left.\begin{aligned} E_y &= Ae^{-i\omega t + ik(x\sin\alpha - z\cos\alpha)} - Ae^{-i\omega t + ik(x\sin\alpha + z\cos\alpha)} \\ &= -2iAe^{-i\omega t + ikx\sin\alpha}\sin(kz\cos\alpha) \\ H_x &= \sqrt{\varepsilon_1}\,2Ae^{-i\omega t + ikx\sin\alpha}\cos\alpha\cos(kz\cos\alpha) \\ H_z &= -\sqrt{\varepsilon_1}\,2iAe^{-i\omega t + ikx\sin\alpha}\sin\alpha\sin(kz\cos\alpha) \end{aligned}\right\}. \qquad [11.1]$$

b. **E** *parallel to the plane of incidence:* $B/A = 1$

$$\left.\begin{aligned} H_x &= -2Ae^{-i\omega t + ikx\sin\alpha}\cos(kz\cos\alpha) \\ E_x &= -\varepsilon_1^{-1/2}\,2iAe^{-i\omega t + ikx\sin\alpha}\cos\alpha\sin(kz\cos\alpha) \\ E_z &= \varepsilon_1^{-1/2}\,2Ae^{-i\omega t + ikx\sin\alpha}\sin\alpha\cos(kz\cos\alpha) \end{aligned}\right\}. \qquad [11.2]$$

Normal incidence. Both cases are the same:

$$\left.\begin{aligned} E_y &= -2|A|\sin(\omega t + \delta)\sin kz \\ H_x &= \sqrt{\varepsilon_1}\,2|A|\cos(\omega t + \delta)\cos kz \end{aligned}\right\}. \qquad [11.3]$$

Thus

1. $\boldsymbol{E}$ has nodes at $kz = \pi N$, $z = N\lambda/2$. (The first node is at the mirror surface.)
2. $\boldsymbol{H}$ has nodes at $kz = \pi(N + \frac{1}{2})$, $z = \frac{1}{2}(N + \frac{1}{2})\lambda$.

If such a situation is photographed, the first node is found to be at the mirror. Thus, in photometry it is $|\boldsymbol{E}|^2$ that is measured. This can be understood on the basis of the theory of electrons.

Arbitrary incidence. For Case a, from Eq. [11.1] it follows that nodes occur for $kz \cos\alpha = \pi N$ or at $z = \frac{1}{2}N\lambda \cos\alpha$. For Case b, it follows from Eq. [11.2] that for $\alpha \neq 0$ and $\alpha \neq \pi/2$, $|\boldsymbol{E}|^2 = |E_x|^2 + |E_z|^2 \neq 0$, and thus there are no nodes present. From this it can be determined that the plane previously called the "plane of polarization" is parallel to the magnetic vector $\boldsymbol{H}$ (see p. 68). The vector $\boldsymbol{E}$ is perpendicular to this plane.

Chapter 4. Crystal Optics

12. RELATIONS FOR THE WAVE NORMAL

In discussing Maxwell's theory, we have assumed that the medium was isotropic. Now, however, we wish to assume that $\boldsymbol{E}$ is no longer parallel to $\boldsymbol{D}$. Thus, we must write the Maxwell equations in their most general form, where, however, we once again assume a permeability $\mu=1$. It turns out, of course, that at the high frequencies which are applicable to light, $\mu(\omega)$ is close to one. Thus, we have

$$\left.\begin{aligned} \operatorname{curl}\boldsymbol{H} &= \frac{1}{c}\dot{\boldsymbol{D}}, \qquad & \operatorname{div}\boldsymbol{D} &= 0 \\ \operatorname{curl}\boldsymbol{E} &= -\frac{1}{c}\dot{\boldsymbol{H}}, \qquad & \operatorname{div}\boldsymbol{H} &= 0 \end{aligned}\right\}. \tag{12.1}$$

From Eqs. [12.1] there follows

$$\nabla^2\boldsymbol{E} - \operatorname{grad}\operatorname{div}\boldsymbol{E} = \frac{1}{c^2}\ddot{\boldsymbol{D}}, \tag{12.2}$$

where $\operatorname{div}\boldsymbol{E}$ no longer vanishes. For the relation between $\boldsymbol{D}$ and $\boldsymbol{E}$ we must assume

$$\boldsymbol{D} = \boldsymbol{\epsilon}\cdot\boldsymbol{E} \quad \text{or} \quad D_i = \sum_k \varepsilon_{ik}E_k, \tag{12.3}$$

where $\boldsymbol{\epsilon}$ is the dielectric tensor. The Poynting vector becomes

$$\boldsymbol{S} = \frac{c}{4\pi}(\boldsymbol{E}\times\boldsymbol{H})$$

and, from the law of conservation of energy, there follows (cf. Eq. [8.6])

$$\frac{\partial W}{\partial t} = -\operatorname{div} \boldsymbol{S} = -\frac{c}{4\pi}(\boldsymbol{H}\cdot\operatorname{curl}\boldsymbol{E} - \boldsymbol{E}\cdot\operatorname{curl}\boldsymbol{H}),$$

where W is the energy density of the field. From this, with Eqs. [12.1],

$$\frac{\partial W}{\partial t} = \frac{1}{8\pi}\frac{\partial}{\partial t}\boldsymbol{H}^2 + \frac{1}{4\pi}\boldsymbol{E}\cdot\frac{\partial \boldsymbol{D}}{\partial t}. \tag{12.4}$$

This must, of course, be an exact differential, so that we must postulate

$$\boldsymbol{E}\cdot\frac{\partial \boldsymbol{D}}{\partial t} = \boldsymbol{D}\cdot\frac{\partial \boldsymbol{E}}{\partial t}, \tag{12.5}$$

for then

$$\frac{1}{4\pi}\boldsymbol{E}\cdot\frac{\partial \boldsymbol{D}}{\partial t} = \frac{1}{8\pi}\frac{\partial}{\partial t}(\boldsymbol{E}\cdot\boldsymbol{D}).$$

If Eq. [12.3] is substituted into Eq. [12.5], there results

$$\sum_{ik} \varepsilon_{ik} E_i \dot{E}_k = \sum_{lm} \varepsilon_{lm} E_m \dot{E}_l,$$

or

$$\sum_{ik} (\varepsilon_{ik} - \varepsilon_{ki}) E_i \dot{E}_k = 0,$$

$$\varepsilon_{ik} = \varepsilon_{ki}. \tag{12.6}$$

Thus, the law of conservation of energy requires that the tensor $\boldsymbol{\epsilon}$ be symmetric. However, such a tensor can always be transformed to principal axes. The characteristic equation for the eigenvalues is

$$\begin{vmatrix} \varepsilon_{11} - \lambda & \varepsilon_{12} & \varepsilon_{13} \\ \varepsilon_{21} & \varepsilon_{22} - \lambda & \varepsilon_{23} \\ \varepsilon_{31} & \varepsilon_{32} & \varepsilon_{33} - \lambda \end{vmatrix} = 0. \tag{12.7}$$

Furthermore, we know that the directions of the principal axes corresponding to two *different* roots of the characteristic equation are perpendicular to one another. If two roots are the same, then the corresponding directions of the principal axes are undetermined in a plane, so that they can be *chosen* orthogonal. For a (real) symmetric tensor, all eigenvalues are real.

In what follows, we assume that the transformation to principal axes has been made at each point of the crystal, so that the tensor $\boldsymbol{\epsilon}$ is diagonal and can be written as follows:

$$(\varepsilon_{ik}) = \begin{pmatrix} \varepsilon_1 & 0 & 0 \\ 0 & \varepsilon_2 & 0 \\ 0 & 0 & \varepsilon_3 \end{pmatrix}. \qquad [12.8]$$

The ε_i are then called the *principal dielectric constants*. We now consider those solutions of Maxwell's equations which can be represented by plane waves. Thus, all field intensities are to propagate as plane waves:

$$\left.\begin{array}{l} \boldsymbol{E} = \boldsymbol{E}_0 e^{i(\boldsymbol{k}\cdot\boldsymbol{x}-\omega t)}, \quad \boldsymbol{H} = \boldsymbol{H}_0 e^{i(\boldsymbol{k}\cdot\boldsymbol{x}-\omega t)} \\ \boldsymbol{k} = \boldsymbol{n}\dfrac{\omega}{v} = \boldsymbol{p}\dfrac{\omega}{c} \\ \boldsymbol{n} = \text{unit vector in the direction of propagation} \\ v = \text{phase velocity} \\ p^2 = \dfrac{c^2}{v^2} = n^2, \quad n = \text{index of refraction} \end{array}\right\}. \qquad [12.9]$$

The vector $\boldsymbol{n}$ is also called the wave normal (since it is perpendicular to the surfaces of constant phase). We have

$$\boldsymbol{p} = n\boldsymbol{n}.$$

We now substitute the forms given in [12.9] into Maxwell's equations [12.1]. There follows

$$\frac{\partial}{\partial t} \sim -i\omega, \quad \frac{\partial}{\partial \boldsymbol{x}} \sim i\boldsymbol{k} = \frac{i\omega}{c}\boldsymbol{p}, \qquad [12.10]$$

$$\boldsymbol{D} = -\boldsymbol{p}\times\boldsymbol{H}, \qquad \boldsymbol{H} = \boldsymbol{p}\times\boldsymbol{E}, \qquad [12.11]$$

$$\left.\begin{array}{r} \boldsymbol{D} = p^2\boldsymbol{E} - \boldsymbol{p}(\boldsymbol{p}\cdot\boldsymbol{E}) \\ \dfrac{1}{n^2}\boldsymbol{D} = \boldsymbol{E} - \boldsymbol{n}(\boldsymbol{n}\cdot\boldsymbol{E}) \end{array}\right\}. \qquad [12.12]$$

From Eq. [12.11] we conclude that $\boldsymbol{D}$ and $\boldsymbol{H}$ are perpendicular to $\boldsymbol{p}$, and that $\boldsymbol{D}$, $\boldsymbol{E}$, and $\boldsymbol{p}$ lie in a plane.

Equation [12.12] together with Eq. [12.3] yields

$$\left(\frac{1}{n^2}-\frac{1}{\varepsilon_i}\right)D_i+n_i\left(\sum_k \frac{D_k}{\varepsilon_k}n_k\right)=0\,. \qquad [12.13]$$

If we set

$$\lambda=\sum_k \frac{n_k}{\varepsilon_k}D_k\,, \qquad [12.14]$$

then Eq. [12.13] becomes

$$\left(\frac{1}{n^2}-\frac{1}{\varepsilon_i}\right)D_i+\lambda n_i=0\,. \qquad [12.15]$$

Equation [12.15] is equivalent to [12.13] if we require that λ be taken so as to satisfy the requirement

$$\sum_k D_k n_k=0\,, \qquad [12.15a]$$

which follows from $\operatorname{div}\boldsymbol{D}=0$ and Eq. [12.10]. (This condition follows directly by multiplying Eq. [12.15] by n_i and then summing.) Thus, instead of [12.13] we can use the system of Eqs. [12.15] and [12.15a].

We solve Eq. [12.15] for D_i,

$$D_i=-\frac{\lambda n_i}{1/n^2-1/\varepsilon_i}=-\frac{\lambda n p_i}{1-p^2/\varepsilon_i}\,, \qquad [12.16]$$

and then take the scalar product with $\boldsymbol{p}$. There follows

$$2G=\sum_i \frac{n_i^2}{1/n^2-1/\varepsilon_i}=\sum_i \frac{p_i^2}{1-p^2/\varepsilon_i}=0\,. \qquad [12.17]$$

Written differently,

$$2G=\sum_i \frac{p_i^2-p^2p_i^2/\varepsilon_i+p^2p_i^2/\varepsilon_i}{1-p^2/\varepsilon_i}=\sum_i\left(p_i^2+p^2\frac{p_i^2/\varepsilon_i}{1-p^2/\varepsilon_i}\right)$$

$$=p^2+p^2\sum_i \frac{p_i^2/\varepsilon_i}{1-p^2/\varepsilon_i}=0\,.$$

Thus,

$$\sum_i \frac{p_i^2/\varepsilon_i}{p^2/\varepsilon_i - 1} = \sum_i \frac{p_i^2}{p^2 - \varepsilon_i} = 1 . \qquad [12.18]$$

If we clear the denominator in Eq. [12.17], then there follows

$$n_1^2\left(\frac{1}{n^2} - \frac{1}{\varepsilon_2}\right)\left(\frac{1}{n^2} - \frac{1}{\varepsilon_3}\right) + n_2^2\left(\frac{1}{n^2} - \frac{1}{\varepsilon_3}\right)\left(\frac{1}{n^2} - \frac{1}{\varepsilon_1}\right) + n_3^2\left(\frac{1}{n^2} - \frac{1}{\varepsilon_1}\right)\left(\frac{1}{n^2} - \frac{1}{\varepsilon_2}\right) = 0 . \qquad [12.19]$$

This yields a quadratic equation for $1/n^2$. Thus, in general, we obtain two different roots for the refractive index. To each wave normal then correspond two different phase velocities.

Special case: $\varepsilon_2 = \varepsilon_3$

Form the preceding discussion (see the remark to Eq. [12.7]), this implies an optical rotational symmetry Equation [12.19] splits into two factors:

$$\left(\frac{1}{n^2} - \frac{1}{\varepsilon_2}\right)\left\{n_1^2\left(\frac{1}{n^2} - \frac{1}{\varepsilon_2}\right) + (n_2^2 + n_3^2)\left(\frac{1}{n^2} - \frac{1}{\varepsilon_1}\right)\right\} = 0 . \qquad [12.20]$$

Case 1. *The first factor is zero*: $1/n^2 = 1/\varepsilon_2$.

The corresponding index of refraction is independent of the wave direction $\boldsymbol{n}$. This is the so-called *ordinary ray*. If we denote the plane through axes 2 and 3 as the principal section, then we can say that the $\boldsymbol{D}$ corresponding to the ordinary ray must lie in the principal section and be perpendicular to $\boldsymbol{n}$. This is true because in Eq. [12.15] we must have $\lambda = 0$ (we set $i = 2$); for $i = 1$, however, $(1/n^2 - 1/\varepsilon_1) \neq 0$, so that $D_1 = 0$. Thus, for the ordinary ray $\boldsymbol{E}$ is parallel to $\boldsymbol{D}$ (because the latter is in the principal section). From Eq. [12.11], the magnetic vector for the ordinary ray then becomes

$$\boldsymbol{H} = n(\boldsymbol{n} \times \boldsymbol{E}) = \frac{1}{\sqrt{\varepsilon_2}} \boldsymbol{n} \times \boldsymbol{D} . \qquad [12.21]$$

If $\boldsymbol{n}$ is perpendicular to the principal section (i.e., parallel to the 1 axis), then the direction of $\boldsymbol{D}$ in the principal plane (and thus the direction of the ordinary ray) is not defined.

Case 2. The second factor in Eq. [12.20] is zero:

$$n_1^2\left(\frac{1}{n^2}-\frac{1}{\varepsilon_2}\right)+(n_2^2+n_3^2)\left(\frac{1}{n^2}-\frac{1}{\varepsilon_1}\right)=0 .$$

From this (since $\boldsymbol{n}^2=1$), there follows

$$\frac{1}{n^2}=\frac{n_1^2}{\varepsilon_2}+\frac{n_2^2+n_3^2}{\varepsilon_1} . \qquad [12.22]$$

If we denote the angle between $\boldsymbol{n}$ and the 1 axis by φ, then there follows from Eq. [12.22]

$$\frac{\cos^2\varphi}{\varepsilon_2}+\frac{\sin^2\varphi}{\varepsilon_1}=\frac{1}{n^2} . \qquad [12.23]$$

Here, the refractive index is dependent on the direction of the normal. This is the *extraordinary ray.*

We consider an extraordinary ray lying in the 1-2 plane (the 2 and 3 axes are suitably chosen), so that $n_3=0$. Then, from Eq. [12.15] it follows that

$$\left(\frac{1}{n^2}-\frac{1}{\varepsilon_3}\right)D_3=0 \quad \text{and therefore} \quad D_3=0 .$$

Thus, the $\boldsymbol{D}$ vector lies in the 1-2 plane and is perpendicular to $\boldsymbol{n}$. If we consider an analogous ordinary ray, then we know that its $\boldsymbol{D}$ vector lies in the principal section and is perpendicular to $\boldsymbol{n}$. Thus, we conclude that the ordinary and extraordinary rays are polarized perpendicular to one another. With the use of [12.21], this can also be expressed by saying that $\boldsymbol{H}$ of the ordinary ray is in the same direction as $\boldsymbol{D}$ of the extraordinary ray.

If in Eq. [12.23] we set $\varphi=0$, then the ordinary ray can no longer be differentiated from the extraordinary ray.

Thus, in our special case there is one (and only one) direction for which Eq. [12.19] has a double root for $1/n^2$. Hence, crystals of this kind are called *optically uniaxial*, and the direction for which there is the double root is the optical axis.

Geometrical interpretation of the general case.

We set

$$\boldsymbol{x} = \text{constant} \times \boldsymbol{D} . \qquad [12.24]$$

From Eq. [12.16], then,

$$\boldsymbol{x} \cdot \boldsymbol{n} = 0 . \qquad [12.25]$$

If we take the ellipsoid

$$\sum_i \frac{x_i^2}{\varepsilon_i} = 1 \qquad [12.26]$$

and intersect it with the wave plane [12.25], then the principal axes of the ellipse formed by this intersection are proportional to the squares of the refractive indices n_1 and n_2 which belong to those polarizations of the plane waves of wave normal $\boldsymbol{n}$ whose vectors $\boldsymbol{D}$ fall along the principal axes under consideration. This statement is the same as saying that the squares of the refractive indices n_1 and n_2 are proportional to the extrema of the quadratic form

$$\sum_i x_i^2 = \text{extremum} \qquad [12.27]$$

under the requirement of the auxiliary conditions [12.25] and [12.26]. With suitably defined constants, the method of Lagrangian multipliers yields

$$\frac{x_i}{\varepsilon_i} - \mu x_i + \gamma n_i = 0 . \qquad [12.28]$$

If we take the scalar product of this with x_i, then upon using Eq. [12.25] and [12.26] there follows for the extreme value of [12.27]

$$\left(\sum_i x_i^2\right)_{\text{extr}} = \frac{1}{\mu}\,. \qquad [12.29]$$

On the other hand, $\boldsymbol{D}$ satisfies Eq. [12.15] and is given by Eq. [12.24]. Substituting the latter as a geometric fact into the former yields the ellipsoid

$$\frac{x_i}{n^2} - \frac{x_i}{\varepsilon_i} + \lambda n_i = 0\,, \qquad [12.30]$$

which is compatible with Eq. [12.28] only if $\mu = 1/n^2$, so that

$$\left(\sum_i x_i^2\right)_{\text{extr}} = \text{constant} \times n^2\,. \qquad \text{Q.E.D.}$$

13. RAY VARIABLES

a. The ray index

Instead of the wave variables used previously, we can also introduce ray variables. To define these, we proceed as follows. Equations [12.11] are relations between $\boldsymbol{p}$, $\boldsymbol{E}$, $\boldsymbol{D}$, and $\boldsymbol{H}$. For the Poynting vector we had

$$\boldsymbol{S} = \frac{c}{4\pi}\,\boldsymbol{E}\times\boldsymbol{H}\,, \qquad W = \frac{1}{8\pi}\,(\boldsymbol{E}\cdot\boldsymbol{D} + |\boldsymbol{H}|^2)\,. \qquad [13.1]$$

We now define the *ray velocity* $\boldsymbol{U}$ by

$$\boldsymbol{U} = \frac{\boldsymbol{S}}{W} = \frac{c}{s}\,\boldsymbol{t}\,. \qquad [13.2]$$

The second part of Eq. [13.2] simultaneously defines the *ray index* s and the *ray vector* $\boldsymbol{t}$. (We have assumed $|\boldsymbol{t}|^2 = 1$.) For isotropic media, the ray index s is equal to the index of refraction n, and the ray vector $\boldsymbol{t}$ is equal to the wave normal vector $\boldsymbol{n}$. For crystals, however, this is no longer the case. A relation between the ray index and the refractive index is obtained as follows:

$$\frac{c}{s}\boldsymbol{t} = \frac{\boldsymbol{S}}{W} = \frac{c\boldsymbol{E}\times\boldsymbol{H}}{\frac{1}{2}\{\boldsymbol{E}\cdot\boldsymbol{D}+|\boldsymbol{H}|^2\}} = \frac{c\boldsymbol{E}\times\boldsymbol{H}}{|\boldsymbol{H}|^2},$$

since for a plane wave the electric energy density is equal to the magnetic energy density. Indeed, from Eq. [12.11], there follows

$$\boldsymbol{D}\cdot\boldsymbol{E} = -(\boldsymbol{p}\times\boldsymbol{H})\cdot\boldsymbol{E} = (\boldsymbol{p}\times\boldsymbol{E})\cdot\boldsymbol{H} = |\boldsymbol{H}|^2.$$

Thus, from Eq. [12.11] and [13.2] there follows

$$\frac{c}{s}\boldsymbol{t} = \frac{c\boldsymbol{E}\times\boldsymbol{H}}{\boldsymbol{H}\cdot(\boldsymbol{p}\times\boldsymbol{E})} = -\frac{c\boldsymbol{E}\times\boldsymbol{H}}{\boldsymbol{p}\cdot(\boldsymbol{H}\times\boldsymbol{E})} = \frac{c\boldsymbol{t}}{\boldsymbol{p}\cdot\boldsymbol{t}},$$

from which $\boldsymbol{p}\cdot\boldsymbol{t} = \boldsymbol{s}$, or

$$\boldsymbol{n}\cdot\boldsymbol{t} = \frac{s}{n}. \qquad [13.3]$$

b. Duality

If, in addition to the above, we also introduce the variable

$$\boldsymbol{q}' = \frac{1}{s}\boldsymbol{t}, \qquad [13.4]$$

then we can convince ourselves that all equations in this chapter remain correct if the following substitution of the first by the second line is made:

$$\begin{matrix} \boldsymbol{D} & \boldsymbol{p} & n & \varepsilon_k & \boldsymbol{n} & \boldsymbol{H} \\ \boldsymbol{E} & \boldsymbol{q}' & 1/s & 1/\varepsilon_k & \boldsymbol{t} & \boldsymbol{H} \end{matrix}. \qquad [13.5]$$

The variables in each column are said to be *duals* of one another.

The substitution [13.5] produces a correct equation out of *each* equation, provided that the original equation is correct. Thus, we must show that Eq. [12.11] permits this substitution:

$$\boldsymbol{H}\times\boldsymbol{S} = \frac{c}{4\pi}\boldsymbol{H}\times(\boldsymbol{E}\times\boldsymbol{H}) = \frac{c}{4\pi}\boldsymbol{E}|\boldsymbol{H}|^2 - \frac{c}{4\pi}\boldsymbol{H}(\boldsymbol{E}\cdot\boldsymbol{H}) = cW\boldsymbol{E},$$

and, analogously, $\boldsymbol{D}\times\boldsymbol{S}=-cW\boldsymbol{H}$. However, with $\boldsymbol{S}=cW\boldsymbol{q}'$,

$$\boldsymbol{E}=-\boldsymbol{q}'\times\boldsymbol{H}, \qquad \boldsymbol{H}=\boldsymbol{q}'\times\boldsymbol{D}, \tag{13.6}$$

which corresponds exactly to Eq. [12.11]. From this, however, everything else follows: from $\boldsymbol{p}$ we obtain $\boldsymbol{q}'$ by substituting $n\to 1/s$, $\boldsymbol{n}\to\boldsymbol{t}$, and the equation (for principal axes) $D_k=\varepsilon_k E_k$ remains correct upon replacing $\boldsymbol{E}\to\boldsymbol{D}$ and $\varepsilon_k\to 1/\varepsilon_k$. Q.E.D.

We know that the direction of the Poynting vector, and thus that of $\boldsymbol{t}$ and $\boldsymbol{q}'$, is the direction of the transport of energy. Thus, $\boldsymbol{q}'$ is the direction in which a wave packet propagates, and $\boldsymbol{n}$ is the direction in which the phase propagates. Thus, all explanations of Section 12 can, with the aid of the scheme of [13.5], be directly taken over for the ray variables.

c. Relation between wave normal and ray direction

We have Eqs. [12.11] and their duals:

$$\boldsymbol{D}=\boldsymbol{H}\times\boldsymbol{p}, \qquad \boldsymbol{E}=\boldsymbol{H}\times\boldsymbol{q}'. \tag{13.7}$$

From these, we obtain

$$\boldsymbol{D}\cdot\boldsymbol{q}'=(\boldsymbol{H}\times\boldsymbol{p})\cdot\boldsymbol{q}'=(\boldsymbol{p}\times\boldsymbol{q}')\cdot\boldsymbol{H}, \tag{13.8}$$

$$\boldsymbol{E}\cdot\boldsymbol{p}=(\boldsymbol{H}\times\boldsymbol{q}')\cdot\boldsymbol{p}=-(\boldsymbol{p}\times\boldsymbol{q}')\cdot\boldsymbol{H}, \tag{13.9}$$

so that

$$\boldsymbol{p}\cdot\boldsymbol{E}=-\boldsymbol{q}'\cdot\boldsymbol{D}. \tag{13.10}$$

On the other hand, we have Eq. [12.12],

$$\boldsymbol{D}=p^2\boldsymbol{E}-\boldsymbol{p}(\boldsymbol{p}\cdot\boldsymbol{E}) \tag{13.11}$$

and its dual

$$\boldsymbol{E}=q'^2\boldsymbol{D}-\boldsymbol{q}'(\boldsymbol{q}'\cdot\boldsymbol{D}). \tag{13.12}$$

In component form (for principal axes),

$$p_i(\boldsymbol{p}\cdot\boldsymbol{E})=p^2E_i-\varepsilon_iE_i, \tag{13.11'}$$

$$-q_i'(\boldsymbol{q}'\cdot\boldsymbol{D})=E_i-q'^2\varepsilon_iE_i. \tag{13.12'}$$

Dividing Eq. [13.11′] by Eq. [13.12′] and using Eq. [13.10] yields

$$\frac{q_i'}{1-\varepsilon_i q'^2}=\frac{p_i}{p^2-\varepsilon_i}\,. \qquad [13.13]$$

Furthermore, we have Eq. [12.17],

$$\sum_i \frac{p_i^2}{1-p^2/\varepsilon_i}=0\,,$$

whose dual is

$$\sum_i \frac{q_i'^2}{1-\varepsilon_i q'^2}=0\,, \qquad [13.14]$$

and Eq. [13.3], which can be transformed to

$$\sum_i p_i q_i'=1\,. \qquad [13.15]$$

From Eq. [13.13] there follows [we wish to calculate $q_i'=q_i'(p)$]

$$q_i'=\left(q'^2-\frac{1}{\varepsilon_i}\right)\frac{p_i}{1-p^2/\varepsilon_i}=\frac{1}{p^2}\left[(p^2q'^2-1)+\left(1-\frac{p^2}{\varepsilon_i}\right)\right]\frac{p_i}{1-p^2/\varepsilon_i}\,,$$

so that

$$p^2 q_i'=p_i+(p^2q'^2-1)\frac{p_i}{1-p^2/\varepsilon_i}\,, \qquad [13.16]$$

or

$$p^2 q_i'-p_i=(p^2q'^2-1)\frac{p_i}{1-p^2/\varepsilon_i}\,.$$

When taking the vectorial square (that is, squaring the ith component and then summing) and using Eq. [13.15], this gives

$$(p^2)^2q'^2-2p^2+p^2=(p^2q'^2-1)^2\sum_i\frac{p_i^2}{(1-p^2/\varepsilon_i)^2}\,.$$

Thus,

$$p^2=(p^2q'^2-1)\sum_i\frac{p_i^2}{(1-p^2/\varepsilon_i)^2}\,. \qquad [13.17]$$

If we set

$$K \equiv \sum_i \frac{p_i^2}{(1-p^2/\varepsilon_i)^2} = K(p)\,, \qquad [13.18]$$

then from Eq. [13.17] there follows

$$p^2 q'^2 - 1 = p^2/K\,,$$

and this when substituted into Eq. [13.16] yields

$$p^2 q_i' = p_i + \frac{p^2}{K}\frac{p_i}{(1-p^2/\varepsilon_i)}\,,$$

so that

$$q_i' = p_i\left[\frac{1}{p^2} + \frac{1}{K}\frac{1}{(1-p^2/\varepsilon_i)}\right]. \qquad [13.19]$$

We can, of course, without calculation write down the corresponding dual equation (using substitution [13.5]):

$$p_i = q_i'\left[\frac{1}{q'^2} + \frac{1}{K'}\frac{1}{(1-q'^2\varepsilon_i)}\right], \qquad [13.20]$$

with

$$K' \equiv \sum_i \left(\frac{q_i'}{1-q'^2\varepsilon_i}\right)^2.$$

At the same time, we see that

$$K'/q'^2 = K/p^2\,. \qquad [13.21]$$

Thus, Eq. [13.13] is solved for q_i' and p_i.

We can represent all of this in still another way by explicitly introducing the function G from Eq. [12.17]:

$$G = \frac{1}{2}\sum_i \frac{p_i^2}{1-p^2/\varepsilon_i} = 0\,. \qquad [13.22]$$

Then, by differentiation we get

$$\frac{\partial G}{\partial p_i} = \frac{p_i}{1-p^2/\varepsilon_i} + p_i \sum_k \frac{(1/\varepsilon_k)p_k^2}{(1-p^2/\varepsilon_k)^2}\,.$$

The second term in this is (see Eq. [13.18])

$$\sum_k \frac{p_k^2/\varepsilon_k}{(1-p^2/\varepsilon_k)^2} = \sum_k \frac{\{(1/\varepsilon_k - 1/p^2) + 1/p^2\} p_k^2}{(1-p^2/\varepsilon_k)^2}$$

$$= \frac{1}{p^2}(-2G + K) = K/p^2 .$$

Thus, we obtain

$$\frac{\partial G}{\partial p_i} = \frac{p_i}{1-p^2/\varepsilon_i} + \frac{1}{p^2} K p_i , \qquad [13.23]$$

and, by comparison with [13.19],

$$\frac{\partial G}{\partial p_i} = K q_i' , \qquad [13.24]$$

$$K = \sum_i p_i \frac{\partial G}{\partial p_i} \qquad [13.25]$$

(in accordance with [13.15]). Following the scheme of [13.5], the corresponding dual equations are

$$F(q') = \frac{1}{2} \sum_i \frac{q_i'^2}{1-\varepsilon_i q'^2} = 0 , \qquad [13.26]$$

$$\frac{\partial F}{\partial q_i'} = K' p_i , \qquad [13.27]$$

$$K' = \sum_i q_i' \frac{\partial F}{\partial q_i'} . \qquad [13.28]$$

d. Geometrical interpretation

1. In Section 12 we have already seen that the refractive indices corresponding to a given wave normal can be obtained from the intersection of an ellipsoid and a plane. In accordance with the scheme of [13.5], the ray indices can also be obtained in this way. If the ellipsoid $\sum_i \varepsilon_i x_i^2 = 1$ is intersected by the plane $\boldsymbol{x} \cdot \boldsymbol{t} = 0$, then the principal axes of the ellipse formed by the intersection are proportional

to the reciprocal of the two ray indices $1/s_1$ and $1/s_2$ that correspond to the ray direction $\boldsymbol{t}$.

2. Equations [13.22] to [13.25] and their duals, Eqs. [13.26] to [13.28], also admit a geometrical interpretation. Here, the ray direction corresponding to a wave normal $\boldsymbol{p}$ is obtained by dropping a perpendicular from the origin to the tangent plane which passes through the point of intersection of the wave normal $\boldsymbol{p}$ with the so-called wave surface $G(\boldsymbol{p}) = 0$. This is true because Eq. [13.24] states that the vector $\boldsymbol{q}'$ is perpendicular to this tangent plane. The length R of the perpendicular is (see Eq. [13.15])

$$R = \boldsymbol{p} \cdot \boldsymbol{t} = s(\boldsymbol{p} \cdot \boldsymbol{q}') = s\,. \qquad [13.29]$$

Thus, the length is exactly equal to the ray index. The pedal surface of the wave surface $G = 0$ (i.e., the surface

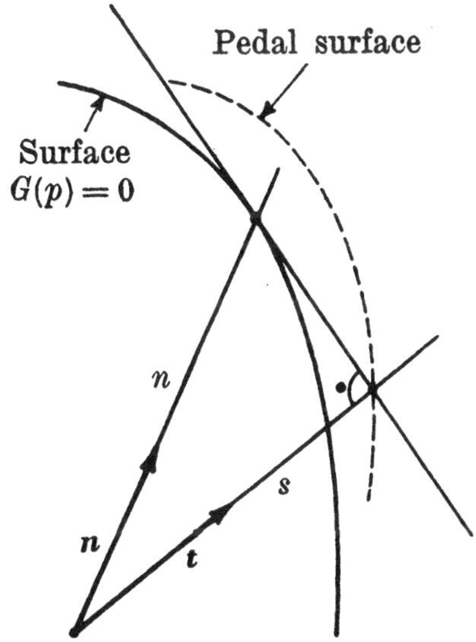

Figure 13.1

consisting of the locus of the points of intersection of the perpendiculars with the planes tangent to the wave surface) is, therefore, the surface obtained by measuring off from the origin a distance s, corresponding to the ray index, along every possible ray direction. Conversely, the surface $G = 0$ is obtained as the envelope of the planes which are constructed perpendicular to radius vectors of length s (equal to the corresponding ray index). The dual constructions are, of course, also valid: the wave normal corresponding to a given ray direction is obtained by dropping a perpendicular from the origin to the tangent plane passing through the point of intersection of the ray vector $\boldsymbol{q}'$ with the so-called ray surface $F(q') = 0$. The length Q of this perpendicular is $1/n$, that is, exactly equal to the reciprocal of the index of refraction. Consequently, the pedal surface of the ray surface $F = 0$ (i.e., the locus of the points of intersection of the perpendiculars with the planes tangent to the ray surface) is the surface obtained by measuring off from the origin a distance $1/n$, the reciprocal of the corresponding refractive index, along every possible wave normal direction. Conversely, the surface $F = 0$ is obtained as the envelope of the planes which are constructed perpendicular to radius vectors of length $1/n$ (n = corresponding index of refraction).

3. There exists a relation between the vector $\boldsymbol{q}'$ and the wave-kinematically defined group velocity:

$$\boldsymbol{U} = c\boldsymbol{q}' . \qquad [13.30]$$

As proof, it must be shown that $\boldsymbol{U}$ satisfies the same equation as $\boldsymbol{q}'$. From our previous work, we have

$$\boldsymbol{k} = \frac{\omega}{c} n\boldsymbol{n} = \frac{\omega}{c}\boldsymbol{p} = k_0\boldsymbol{p} \quad \text{with} \quad k_0 \equiv \frac{\omega}{c}, \quad \text{so that} \quad p_i = \frac{k_i}{k_0} .$$

The group velocity $\boldsymbol{U}$ was defined by

$$\boldsymbol{U} = \frac{\partial \omega}{\partial \boldsymbol{k}}, \qquad \omega = \omega(k_1, k_2, k_3) .$$

From the above equations follows

$$\boldsymbol{U} = c\,\frac{\partial k_0}{\partial \boldsymbol{k}} .$$

For the proof, we introduce the variables k_0, k_1, k_2, and k_3 (homogeneous coordinates) in place of the p_i in the function G:

$$G = \frac{1}{2}\sum_i \frac{p_i^2}{1 - p^2/\varepsilon_i} = \frac{1}{2}\sum_i \frac{k_i^2}{k_0^2 - \boldsymbol{k}^2/\varepsilon_i} \; (=0) . \qquad [13.31]$$

Since Eq. [13.31] defines a homogeneous function of the zeroth degree in k_0, k_1, k_2, k_3, then the Euler relation holds:

$$k_0 \frac{\partial G}{\partial k_0} + \boldsymbol{k} \cdot \frac{\partial G}{\partial \boldsymbol{k}} = 0 . \qquad [13.32]$$

Now, k_0 must be such a function of the k_i that for a variation of the k_i (transition to a neighboring wave normal) the form G must still vanish. This yields the condition

$$\frac{\partial G}{\partial k_0}\frac{\partial k_0}{\partial k_i} + \left(\frac{\partial G}{\partial k_i}\right)_{k_0} = 0 , \qquad \text{or} \qquad \frac{1}{k_0}\frac{\partial G}{\partial p_i} = -\frac{\partial G}{\partial k_0}\frac{\partial k_0}{\partial k_i} .$$

From this, upon using Euler's relation [13.32], it follows further that

$$\frac{\partial G}{\partial p_i} = -k_0 \frac{\partial G}{\partial k_0}\frac{\partial k_0}{\partial k_i} = \left(\boldsymbol{k} \cdot \frac{\partial G}{\partial \boldsymbol{k}}\right)\frac{\partial k_0}{\partial k_i} = \left(\frac{\boldsymbol{k}}{k_0} \cdot \frac{\partial G}{\partial \boldsymbol{p}}\right)\frac{\partial k_0}{\partial k_i} = \left(\boldsymbol{p} \cdot \frac{\partial G}{\partial \boldsymbol{p}}\right)\frac{\partial k_0}{\partial k_i} .$$

Comparing this with Eq. [13.24], we get

$$q_i' = \frac{\partial k_0}{\partial k_i},$$

or

$$q_i' = \frac{1}{c}\frac{\partial \omega}{\partial k_i} = \frac{1}{c} U_i . \qquad \text{Q.E.D.}$$

In concluding these paragraphs we may note that here the quantities $\boldsymbol{p}$, $\boldsymbol{q}'$, and G play roles analogous to those of the similarly designated quantities in geometrical optics (see Section 3b, pp. 16–20).

14. SINGULARITIES

There exist singular directions for which no unique ray direction is associated with the wave normal (and, of course, the same is true for the dual quantities). This is the case when Eq. [12.17] or Eq. [13.31], which determines the index of refraction, has a double root. In that case (in the principal axis system) we have

$$2G = \sum_i \frac{n_i^2}{1/n^2 - 1/\varepsilon_i} = 0 \, .$$

With the abbreviation

$$N = \left(\frac{1}{n^2} - \frac{1}{\varepsilon_1}\right)\left(\frac{1}{n^2} - \frac{1}{\varepsilon_2}\right)\left(\frac{1}{n^2} - \frac{1}{\varepsilon_3}\right),$$

we have

$$2NG = n_1^2\left(\frac{1}{n^2} - \frac{1}{\varepsilon_2}\right)\left(\frac{1}{n^2} - \frac{1}{\varepsilon_3}\right) + n_2^2\left(\frac{1}{n^2} - \frac{1}{\varepsilon_1}\right)\left(\frac{1}{n^2} - \frac{1}{\varepsilon_3}\right) + n_3^2\left(\frac{1}{n^2} - \frac{1}{\varepsilon_1}\right)\left(\frac{1}{n^2} - \frac{1}{\varepsilon_2}\right) = 0 \, .$$

Now, if $NG = 0$ is to have a double root, then it must also be true that $(NG)' = 0$, where the prime denotes differentiation with respect to $1/n^2$. This yields

$$\left.\begin{aligned} NG &= 0 \\ (NG)' &= 0 \end{aligned}\right\} \qquad \text{(definition of a double root)} \, .$$

Hence,

$$NG' = 0 \, .$$

As can easily be shown, G' cannot be zero. Therefore, for a double root in $1/n^2$,

$$N = \left(\frac{1}{n^2} - \frac{1}{\varepsilon_1}\right)\left(\frac{1}{n^2} - \frac{1}{\varepsilon_2}\right)\left(\frac{1}{n^2} - \frac{1}{\varepsilon_3}\right) = 0 \, .$$

This is the case when, for example, $n^2 = \varepsilon_2$. Moreover, all the ε_i are to be different. It then follows that

$$2NG = n_2^2\left(\frac{1}{n^2} - \frac{1}{\varepsilon_1}\right)\left(\frac{1}{n^2} - \frac{1}{\varepsilon_3}\right) = 0 \, ,$$

so that

$$n_2 = 0 \, .$$

If the preceding is substituted into the general formula for NG, then we obtain

$$2NG = \left(\frac{1}{n^2} - \frac{1}{\varepsilon_2}\right)\left\{n_1^2\left(\frac{1}{n^2} - \frac{1}{\varepsilon_3}\right) + n_3^2\left(\frac{1}{n^2} - \frac{1}{\varepsilon_1}\right)\right\}.$$

For a double root, the quantity in curly brackets must also vanish:

$$n_1^2\left(\frac{1}{\varepsilon_2} - \frac{1}{\varepsilon_3}\right) + n_3^2\left(\frac{1}{\varepsilon_2} - \frac{1}{\varepsilon_1}\right) = 0 .$$

This equation can be satisfied only if the two bracketed terms have opposite signs. Thus,

$$\varepsilon_1 < \varepsilon_2 < \varepsilon_3 \quad \text{(or vice versa)}, \qquad [14.1]$$

and for the directions for which the double roots occur there follows

$$\left.\begin{aligned} b_1 &= \pm\sqrt{\frac{1/\varepsilon_1 - 1/\varepsilon_2}{1/\varepsilon_1 - 1/\varepsilon_3}} = \pm\sqrt{\frac{\varepsilon_3(\varepsilon_2 - \varepsilon_1)}{\varepsilon_2(\varepsilon_3 - \varepsilon_1)}} \\ b_2 &= 0 \\ b_3 &= \pm\sqrt{\frac{1/\varepsilon_2 - 1/\varepsilon_3}{1/\varepsilon_1 - 1/\varepsilon_3}} = \pm\sqrt{\frac{\varepsilon_1(\varepsilon_3 - \varepsilon_2)}{\varepsilon_2(\varepsilon_3 - \varepsilon_1)}} \end{aligned}\right\}. \qquad [14.2]$$

Thus, there are two such intrinsically different directions which are called the *binormals* or *optic axes*. In accordance with the general construction, the refractive indices corresponding to a given wave normal are obtained by cutting the ellipsoid $\sum_i x_i^2/\varepsilon_i = 1$ with the plane $\boldsymbol{x}\cdot\boldsymbol{n} = 0$. If $\boldsymbol{n}$ is along a binormal direction, then the intersection is a circle and the binormals pass through its center and through the umbilical points [1] of the ellipse in question. Since the circle has no principal axes, $\boldsymbol{D}$ and, therefore, $\boldsymbol{E}$ are inde-

[1] *Translator's Note*: If we consider a family of parallel planes, each intersecting a quadratic surface in a circle, then the points of tangency of these planes with the surface are called the *umbilical points* of the quadratic surface.

terminate. The only requirement is that $\boldsymbol{D}$ must be perpendicular to the binormals. Now, from Eqs. [12.11] and their duals, we know that $\boldsymbol{E}$, $\boldsymbol{D}$, $\boldsymbol{t}$, and $\boldsymbol{n}$ (in our case $\boldsymbol{n}=\boldsymbol{b}$) lie in one plane. (See Fig. 14.1.) However, $\boldsymbol{D}$ can have

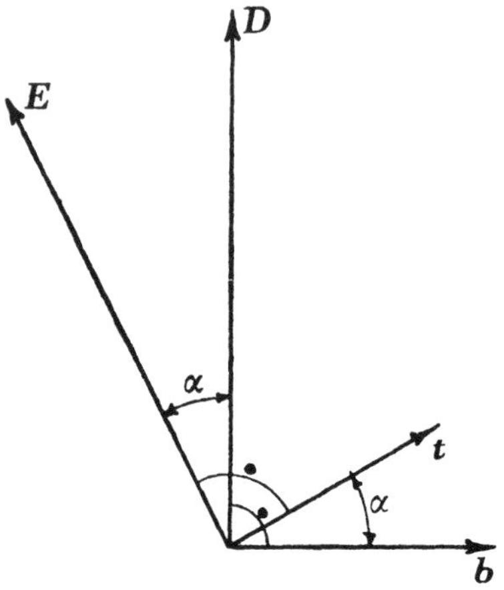

Figure 14.1

any direction perpendicular to $\boldsymbol{b}$, so that the whole figure is undetermined with respect to a rotation about $\boldsymbol{b}$. Thus, we see that $\boldsymbol{t}$ lies on a cone. It should be noted that the angle is dependent upon the azimuth, so that in general $\boldsymbol{t}$ does not lie on a circular cone. Thus, we have proved the hypothesis asserted at the beginning of this section: if the index of refraction is a double root of Eq. [12.17], then no unique ray direction is associated with a given wave normal.

We now wish to derive the equation of this cone. In the principal axes system (which is arbitrary in a plane perpendicular to $\boldsymbol{b}$), $\boldsymbol{D}$ has the following components:

$$\boldsymbol{D}=\{-\lambda b_3,\ \mu,\ \lambda b_1\}, \tag{14.3}$$

for then the scalar product with $\boldsymbol{b}$ is equal to zero. Thus,

$\boldsymbol{E}$ becomes

$$\boldsymbol{E} = \left\{ -\frac{\lambda}{\varepsilon_1} b_3, \frac{\mu}{\varepsilon_2}, \frac{\lambda}{\varepsilon_3} b_1 \right\}. \qquad [14.4]$$

The generatrix of the desired cone has the direction of $\boldsymbol{t}$ and, therefore, is perpendicular to $\boldsymbol{E}$ and lies in the plane containing $\boldsymbol{E}$ and $\boldsymbol{b}$. This leads to the requirements

$$\boldsymbol{b} \cdot (\boldsymbol{E} \times \boldsymbol{x}) = 0, \quad \text{and} \quad \boldsymbol{E} \cdot \boldsymbol{x} = 0,$$

where we have denoted the coordinates, in the principal axes system, of an arbitrary point on the cone by $\boldsymbol{x} = \{\Xi, H, Z\}$. In component form,

$$b_1(E_2 Z - E_3 H) + b_3(E_1 H - E_2 \Xi) = 0,$$

and

$$E_1 \Xi + E_2 H + E_3 Z = 0.$$

Into these we substitute from Eq. [14.4]:

$$b_1 \left(\frac{\mu}{\varepsilon_2} Z - \lambda \frac{b_1}{\varepsilon_3} H \right) + b_3 \left(-\frac{\lambda}{\varepsilon_1} b_3 H - \frac{\mu}{\varepsilon_2} \Xi \right) = 0,$$

and

$$-\frac{\lambda}{\varepsilon_1} b_3 \Xi + \frac{\mu}{\varepsilon_2} H + \frac{\lambda}{\varepsilon_3} b_1 Z = 0.$$

From these, we eliminate λ/μ:

$$(b_1 Z - b_3 \Xi) \frac{\mu}{\varepsilon_2} = \lambda H \left(\frac{b_1^2}{\varepsilon_3} + \frac{b_2^2}{\varepsilon_1} \right),$$

$$H \frac{\mu}{\varepsilon_2} = \lambda \left(\frac{b_3}{\varepsilon_1} \Xi - \frac{b_1}{\varepsilon_3} Z \right),$$

and, by dividing,

$$(b_1 Z - b_3 \Xi) \left(\frac{b_3}{\varepsilon_1} \Xi - \frac{b_1}{\varepsilon_3} Z \right) = H^2 \left(\frac{b_1^2}{\varepsilon_3} + \frac{b_3^2}{\varepsilon_1} \right). \qquad [14.5]$$

Equation [14.5] is clearly that of an elliptical cone. If

the b_i from Eq. [14.2] are substituted, there follows

$$0 = \Xi^2(\varepsilon_3 - \varepsilon_2) + H^2(\varepsilon_3 - \varepsilon_1) + Z^2(\varepsilon_2 - \varepsilon_1) - \Xi Z \sqrt{\frac{(\varepsilon_2 - \varepsilon_1)(\varepsilon_3 - \varepsilon_2)}{\varepsilon_3\varepsilon_1}}\,(\varepsilon_3 + \varepsilon_1)\,.$$

Equation [14.5] can be further transformed as follows, so that the geometrical form of the cone becomes clearer. From Eq. [14.5] there follows

$$H^2\left(\frac{b_1^2}{\varepsilon_3} + \frac{b_3^2}{\varepsilon_1}\right) + Z^2\frac{b_1^2}{\varepsilon_3} + \Xi Z b_1 b_3\left(-\frac{1}{\varepsilon_3} - \frac{1}{\varepsilon_1}\right) + \Xi^2\frac{b_3^2}{\varepsilon_1} = 0\,,$$

$$\left(\frac{b_1^2}{\varepsilon_3} + \frac{b_3^2}{\varepsilon_1}\right)(\Xi^2 + H^2 + Z^2) - \left(Z^2\frac{b_3^2}{\varepsilon_1} + \Xi^2\frac{b_1^2}{\varepsilon_3}\right) + \Xi Z b_1 b_3\left(-\frac{1}{\varepsilon_3} - \frac{1}{\varepsilon_1}\right) = 0\,,$$

$$\left(\frac{b_1^2}{\varepsilon_3} + \frac{b_3^2}{\varepsilon_1}\right)(\Xi^2 + H^2 + Z^2) - (Zb_3 + \Xi b_1)\left(Z\frac{b_3}{\varepsilon_1} + \Xi\frac{b_1}{\varepsilon_3}\right) = 0. \quad [14.6]$$

From Eq. [14.6] we see that the cone defined there has the same curve of intersection with the plane

$$\Xi b_1 + Z b_3 = C \quad [14.7]$$

(that is, the wave plane) as does the sphere

$$\left(\frac{b_1^2}{\varepsilon_3} + \frac{b_3^2}{\varepsilon_1}\right)(\Xi^2 + H^2 + Z^2) - C\left(Z\frac{b_3}{\varepsilon_1} + \Xi\frac{b_1}{\varepsilon_3}\right) = 0$$

with the wave plane [14.7]. We thus have the important theorem that *the planes perpendicular to the binormals intersect the corresponding ray cone in circles.* Furthermore, we know that the cone contains the vector $\boldsymbol{b}$ in its surface (we need only to choose $\boldsymbol{D}$ parallel to the 2 direction), and that the entire geometry is symmetric with respect to the 1-3 plane. The angle α' between $\boldsymbol{b}$ and the other generatrix lying in the 1-3 (symmetry) plane, which is a measure of the opening of the cone (clearly, we can not speak of an opening angle for the case of an elliptical cone), is also the angle between $\boldsymbol{E}$ and $\boldsymbol{D}$ in the 1-3 plane. However, this latter angle can be easily calculated:

$$\begin{aligned}\tan\alpha' &= \tan[(\boldsymbol{E},\boldsymbol{D}) \text{ (in the principal plane)}]\\ &= \pm\tan[(\boldsymbol{E},\mathbf{1})-(\boldsymbol{D},\mathbf{1})]\\ &= [\tan(\boldsymbol{E},\mathbf{1})-\tan(\boldsymbol{D},\mathbf{1})]/[1+\tan(\boldsymbol{E},\mathbf{1})\tan(\boldsymbol{D},\mathbf{1})]\,,\end{aligned}$$

$$\tan(\boldsymbol{D},\mathbf{1}) = D_3/D_1 = -\,b_1/b_3$$

(since $\boldsymbol{D}$ is perpendicular to $\boldsymbol{b}$),

$$\tan(\boldsymbol{E},\mathbf{1}) = E_3/E_1 = D_3\varepsilon_1/(D_1\varepsilon_3) = -\,b_1\varepsilon_1/(b_3\varepsilon_3)\,,$$

so that

$$\pm\tan\alpha' \equiv \pm\tan(\boldsymbol{E},\boldsymbol{D}) = \frac{-\,\varepsilon_1 b_1 b_3 + \varepsilon_3 b_1 b_3}{\varepsilon_3 b_3^2 + \varepsilon_1 b_1^2}\,.$$

If we substitute b_1 and b_3 from Eq. [14.2] and order the terms, there follows

$$\tan\alpha' = \sqrt{\frac{(\varepsilon_3-\varepsilon_2)(\varepsilon_2-\varepsilon_1)}{\varepsilon_1\varepsilon_3}}\,. \qquad [14.8]$$

The entire discussion can be given in a completely analogous way for the dual quantities. The most important theorems are then as follows.

There exist singular directions for which there is no unique wave normal associated with the ray direction. This is the case when the equation that determines the ray index has a double root. The ray directions for which this is true are called *biradials* or *ray axes*, and their direction cosines are

$$r_1 = \pm\sqrt{\frac{\varepsilon_2-\varepsilon_1}{\varepsilon_3-\varepsilon_1}}\,,\qquad r_2 = 0\,,\qquad r_3 = \pm\sqrt{\frac{\varepsilon_3-\varepsilon_2}{\varepsilon_3-\varepsilon_1}}\,.$$

The associated wave normals lie on a cone which intersects the planes perpendicular to the biradials in circles. The angle β' (a measure of the opening of the wave cone), which corresponds to the angle α' discussed above, is given by

$$\tan\beta' = \sqrt{\frac{(\varepsilon_3 - \varepsilon_2)(\varepsilon_2 - \varepsilon_1)}{\varepsilon_2^2}}\,. \qquad [14.8']$$

These results can also be obtained without calculation from the scheme of Eq. [13.5] and the corresponding formulas for the binormals.

15. LIGHT ENTERING AND LEAVING CRYSTALS

Consider the case shown in Fig. 15.1. Here too, we have the law of refraction:

$$\left.\begin{aligned} n_1 \sin\alpha_1 &= n_2 \sin\alpha_2 \\ n_2 &= n_2(\alpha_2) \end{aligned}\right\}. \qquad [15.1]$$

The system of equations [15.1] is very complicated to solve analytically. Huygens, however, gave a construction with

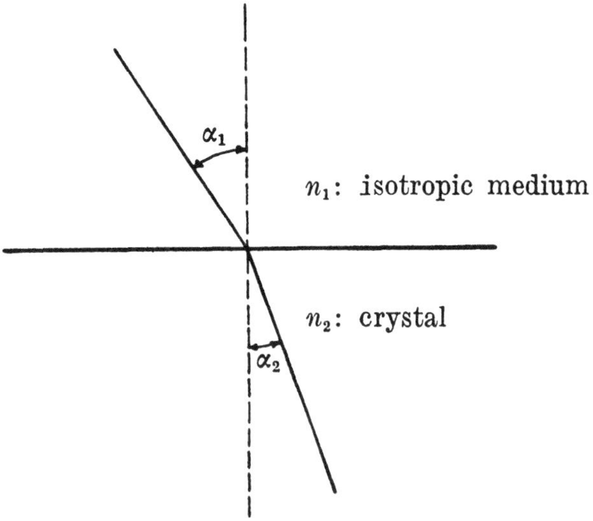

Figure 15.1

which it is easy to get an overall view of the various relationships. Let A_1B_1 be an edge view of part of a surface of constant phase of a plane wave incident on the crystal (the surface of constant phase is perpendicular to the plane of Fig. 15.2). The construction is such that the

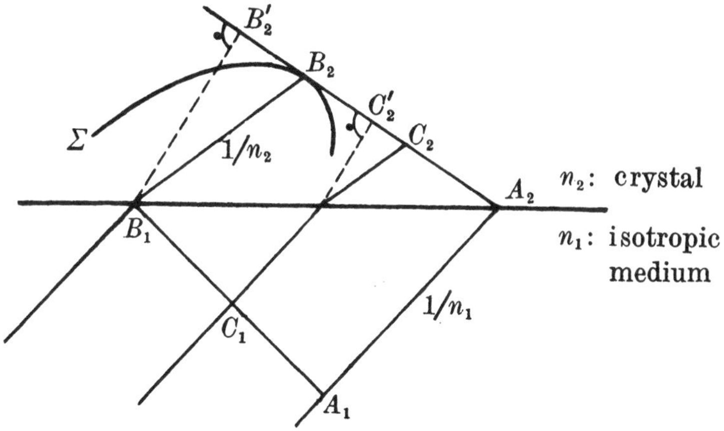

Figure 15.2

optical path from B_1 to B_2 is the same as that from A_1 to A_2; that is, the surface Σ is so defined that the length of each line drawn from B_1 to Σ is the reciprocal of the index of refraction associated with the direction of that line. As can easily be seen, the plane perpendicular to the plane of the drawing which passes through A_2 and is tangent to the surface Σ is a surface of constant phase (the optical path from C_1 to C_2 is always the same, no matter where C_1 is chosen on A_1B_1). Of course, the point of contact does not necessarily lie in the plane of the figure (as it is drawn here for simplicity). Since the wave normal is always perpendicular to the surface of constant phase, it is obtained by dropping a perpendicular, B_1B_2', to the tangent plane mentioned above. The associated ray direction can then be found with the use of a previous construction. It should be noted that the surface Σ consists of two sheets; thus, in the crystal there are two wave normals and two ray directions for one plane wave incident on the crystal. The plane wave is split into two, as is the light ray.[2]

A singularity occurs if the normal direction (B_1B_2') corresponds to a binormal of the crystal. This results in *internal conical refraction.* The refracted rays then lie on a cone (see Section 14). If a light ray is sent through a crystal with two plane parallel surfaces at such an angle that internal conical refraction occurs, then the ray emerges as an elliptical cylinder.

The corresponding dual phenomenon is *external conical refraction.* If a light ray which is parallel to a biradial direction within a crystal emerges from the crystal, then the emerging wave normals lie on a cone produced by the refraction of the indeterminate wave normals which are associated with the biradial ray and which lie on a cone. A suitably chosen convergent pencil of light incident on a crystal plane will, as a result of this phenomenon, form a single ray within the crystal and will emerge from the other side as a divergent pencil [A-5].

[2] *Translator's Note*: This phenomenon is known as *birefringence* or *double refraction.*

Chapter 5. Molecular Optics

16. DISPERSION BY UNDAMPED OSCILLATORS

We now make specific assumptions about the medium in which the optical phenomena take place. In this manner we shall attempt to explain the optical constants.

We assume that charged particles are distributed throughout the medium under consideration. Usually it will suffice to replace the discrete distribution by a continuous one. If

$$\sum_k e_k = 0\,, \qquad [16.1]$$

in which $\sum_k$ is the sum over all particles in a unit volume, and e_k is the charge of the kth particle, then we denote

$$\boldsymbol{P} = \sum_k e_k \boldsymbol{x}_k \qquad [16.2]$$

as the electric dipole moment per unit volume. The current density is

$$\boldsymbol{i} = \dot{\boldsymbol{P}} = \sum_k e_k \boldsymbol{v}_k\,. \qquad [16.3]$$

For many purposes it may be assumed that the positive and negative particles are of the same type, so that we can write

$$\boldsymbol{P} = N_+ e_+ \boldsymbol{x}_+ + N_- e_- \boldsymbol{x}_-\,, \qquad [16.4]$$

where $N_\pm$, $e_\pm$, $\boldsymbol{x}_\pm$ represent the number of particles per unit volume, the charge, and the position of the center of charge, respectively, of the two kinds of particles.

For this kind of a medium the Maxwell equations are

$$\left.\begin{aligned} \operatorname{curl} \boldsymbol{E} &= -\frac{1}{c}\dot{\boldsymbol{H}}, \qquad \operatorname{div} \boldsymbol{H} = 0 \\ \operatorname{curl} \boldsymbol{H} &= \frac{1}{c}\dot{\boldsymbol{D}}, \qquad \operatorname{div} \boldsymbol{D} = 0 \end{aligned}\right\}. \tag{16.5}$$

To these must be added the relation between $\boldsymbol{D}$ and $\boldsymbol{E}$:

$$\boldsymbol{D} = \boldsymbol{E} + 4\pi \boldsymbol{P}. \tag{16.6}$$

Here, as before, we have assumed $\mu = 1$ (no magnetization). For a wave field periodic with the time factor $e^{-i\omega t}$, we have already seen that $\boldsymbol{D} = n^2 \boldsymbol{E}$, so that with Eq. [16.6] there follows

$$4\pi \boldsymbol{P} = (n^2 - 1)\boldsymbol{E}. \tag{16.7}$$

Then, the Maxwell equations have as solutions waves with phase velocity c/n.

Furthermore, we will assume that the charged particles of the medium behave as quasi-elastic oscillators with respect to external influences. Then, the restoring force on a particle is

$$\boldsymbol{K} = -f\boldsymbol{x},$$

where $\boldsymbol{x}$ is the displacement from the equilibrium position. Without external forces, the differential equation of motion for the particle is

$$m\ddot{\boldsymbol{x}} + f\boldsymbol{x} = 0, \qquad m\omega_0^2 = f. \tag{16.8}$$

This assumption is consistent with the occurrence of sharp spectral lines.

Now, assume that an electric field acts on the medium. This gives rise to an additional force

$$\boldsymbol{K}_A = e\boldsymbol{E} \tag{16.9}$$

on a particle, so that

$$m(\ddot{\boldsymbol{x}} + \omega_0^2 \boldsymbol{x}) = e\boldsymbol{E}. \tag{16.10}$$

If the electric field is periodic, that is, if

$$\boldsymbol{E} = \boldsymbol{E}_0 e^{-i\omega t}, \tag{16.11}$$

then Eq. [16.10] admits a solution of the form

$$\boldsymbol{x} = \boldsymbol{x}_0 e^{-i\omega t}. \tag{16.12}$$

Substituting Eq. [16.12] into Eq. [16.10] yields

$$m(-\omega^2 + \omega_0^2)\,\boldsymbol{x}_0 = e\boldsymbol{E}_0, \tag{16.12'}$$

so that

$$\boldsymbol{x}_0 = \frac{e}{m}\,\frac{1}{(\omega_0^2 - \omega^2)}\,\boldsymbol{E}_0 \tag{16.13}$$

and

$$\boldsymbol{x} = \frac{e}{m}\,\frac{1}{(\omega_0^2 - \omega^2)}\,\boldsymbol{E}. \tag{16.14}$$

From this, we obtain a formula for the electric polarization and the index of refraction. If N_0 is the number of particles per unit volume having a free oscillation frequency ω_0, then there follows (if there is no polarization in the absence of the electric field $\boldsymbol{E}$)

$$\boldsymbol{P} = N_0\frac{e^2}{m}\,\frac{1}{(\omega_0^2 - \omega^2)}\,\boldsymbol{E}, \tag{16.15}$$

and from Eq. [16.7]

$$n^2 - 1 = 4\pi N_0\,\frac{e^2}{m}\,\frac{1}{(\omega_0^2 - \omega^2)}. \tag{16.16}$$

If there are several types of oscillators, say N_k of frequency ω_k per unit volume, then there follows

$$n^2 - 1 = 4\pi \sum_k N_k \frac{e_k^2}{m_k}\,\frac{1}{(\omega_k^2 - \omega^2)}. \tag{16.16'}$$

Here, of course, we have made the essential assumption that the number of particles within a cube the length of whose side is a wavelength is large. This is no longer correct for the case of x rays.

Equation [16.16′] is only conditionally correct because the field intensity $\boldsymbol{E}$ is not equal to the field intensity $\boldsymbol{E}'$ that produces the polarization. The field intensity in Eq. [16.10] is understood to be the exciting field, while that in Eq. [16.7] is understood to be the average field. The connection between $\boldsymbol{E}$ and $\boldsymbol{E}'$ is determined in electrodynamics in the derivation of the Clausius-Mosotti equation. We have

$$\boldsymbol{E} = \boldsymbol{E}' - \varphi \boldsymbol{P}, \qquad [16.17]$$

where for isotropic bodies

$$\varphi = 4\pi/3 . \qquad [16.17']$$

With this correction, there follows

$$\left.\begin{aligned} 4\pi \boldsymbol{P} &= (n^2-1)\boldsymbol{E}' - \frac{4\pi}{3}(n^2-1)\boldsymbol{P} \\ \text{or} \quad \frac{4\pi}{3}(n^2+2)\boldsymbol{P} &= (n^2-1)\boldsymbol{E}' \end{aligned}\right\} . \qquad [16.18]$$

This, now, is the equation which must be substituted into Eq. [16.15]. There then follows

$$3\frac{n^2-1}{n^2+2} = 4\pi N_0 \frac{e^2}{m}\frac{1}{(\omega_0^2-\omega^2)}, \qquad [16.19]$$

or, if we again allow several kinds of particles,

$$3\frac{n^2-1}{n^2+2} = 4\pi \sum_k N_k \frac{e_k^2}{m_k}\frac{1}{(\omega_k^2-\omega^2)} . \qquad [16.19']$$

Of course, even this equation is not correct since the right-hand side of Eq. [16.19] has a pole at $\omega = \omega_k$. Thus, we will have to assume a damping force later. For the present, we wish to pursue somewhat further the results that follow from this model of quasi-elastic, charged particles.

a. Molar refractivity

1. Let the moving particles be electrons, let M be the molecular weight of the substance under consideration, and let ϱ be the density. We define "molar refractivity" R by

$$R = \frac{n^2 - 1}{n^2 + 2} \frac{M}{\varrho} . \qquad [16.20]$$

Then, there follows from Eq. [16.19′],

$$R = \frac{4\pi}{3} A \sum_k \frac{e_k^2}{m_k} \frac{f_k}{(\omega_k^2 - \omega^2)} . \qquad [16.21]$$

Here, A is Avogadro's number, and f_k the number of oscillators per actual molecule. According to our model, it would be expected that the f_k are integers (since they are considered as being particles that oscillate). In reality, even for the hydrogen atom (a single electron) there are many such "oscillators," and only the sum of all these "numbers" of oscillators is equal to the number of electrons Z in the molecule:

$$\sum_k f_k = Z . \qquad [16.22]$$

Thus, f_k is called the *oscillator strength* corresponding to the frequency ω_k [A-6].

For the limiting case $\omega \gg \omega_k$ for all k, there follows from Eqs. [16.21] and [16.22]

$$R \sim \frac{4\pi}{3} A \frac{e^2}{m\omega^2} Z .$$

(Here, e and m are constants of the electron.) In this limiting case, the electrons are considered as being free.

2. Let the moving particles be atomic nuclei. For oscillations of diatomic molecules we have $f = 1$ and $1/m = 1/M_1 + 1/M_2$, where M_1 and M_2 are the masses of the nuclei. Since the oscillator strength is an integer, the model of oscillating particles does not break down as it did in 1.

b. Conservation of energy

We again neglect the difference between $\boldsymbol{E}'$ and $\boldsymbol{E}$. The Poynting vector is

$$\boldsymbol{S} = \frac{c}{4\pi} \boldsymbol{E} \times \boldsymbol{H},$$

so that

$$\operatorname{div} \boldsymbol{S} = \frac{c}{4\pi} (\boldsymbol{H} \cdot \operatorname{curl} \boldsymbol{E} - \boldsymbol{E} \cdot \operatorname{curl} \boldsymbol{H}).$$

However, since we have the familiar energy equation

$$\operatorname{div} \boldsymbol{S} + \dot{W} = 0,$$

then if we use Maxwell's equations [16.5] and [16.6] there follows

$$\dot{W} = \frac{1}{4\pi} (\boldsymbol{H} \cdot \dot{\boldsymbol{H}} + \boldsymbol{E} \cdot \dot{\boldsymbol{E}}) + \boldsymbol{E} \cdot \dot{\boldsymbol{P}}.$$

However, in a unit volume we have

$$\boldsymbol{P} = \sum_k N_k e_k \boldsymbol{x}_k$$

and

$$\ddot{\boldsymbol{x}}_k + \omega_k^2 \boldsymbol{x}_k = \frac{e_k}{m_k} \boldsymbol{E},$$

so that there follows

$$\begin{aligned} \boldsymbol{E} \cdot \dot{\boldsymbol{P}} &= \sum_k N_k m_k \dot{\boldsymbol{x}}_k \cdot (\ddot{\boldsymbol{x}}_k + \omega_k^2 \boldsymbol{x}_k) \\ &= \frac{\mathrm{d}}{\mathrm{d}t} \left\{ \sum_k N_k \frac{m_k}{2} (\dot{\boldsymbol{x}}_k^2 + \omega_k^2 \boldsymbol{x}_k^2) \right\}. \end{aligned}$$

Thus, after integration,

$$W = \frac{1}{8\pi} (\boldsymbol{H}^2 + \boldsymbol{E}^2) + \sum_k N_k \frac{m_k}{2} (\dot{\boldsymbol{x}}_k^2 + \omega_k^2 \boldsymbol{x}_k^2),$$

$$W_{\text{total}} = W_{\text{el. mag.}} + W_{\text{material oscillators}},$$

which is the law of conservation of energy.

Here too we can show that

$$\frac{\boldsymbol{S}}{W}=\boldsymbol{U}=\frac{\partial\omega}{\partial\boldsymbol{k}}$$

is the group velocity. In order to see this, we first calculate the energy density of the material oscillators. We have just shown that this is

$$W_{\text{mat}}=\sum_k N_k\frac{m_k}{2}\left(\dot{\boldsymbol{x}}_k^2+\omega_k^2\boldsymbol{x}_k^2\right).$$

Now, however, we have Eq. [16.14], so that

$$\boldsymbol{x}_k=\frac{(e/m)_k}{\omega_k^2-\omega^2}\boldsymbol{E}=\frac{(e/m)_k}{\omega_k^2-\omega^2}\operatorname{Re}\boldsymbol{E}_0e^{-i\omega t-i\delta}. \qquad [16.14']$$

Since we are dealing with quadratic expressions, we must first take the real parts of the complex quantities. When Eq. [16.14′] is substituted into the relation for W_{mat}, there follows

$$\begin{aligned}W_{\text{mat}}&=\sum_k N_k\frac{m_k}{2}\frac{(e/m)_k^2}{(\omega_k^2-\omega^2)^2}\\&\qquad\times\left\{\left(\operatorname{Re}\left(-i\omega\boldsymbol{E}_0e^{-i\omega t-i\delta}\right)\right)^2+\omega_k^2\left(\operatorname{Re}\boldsymbol{E}_0e^{-i\omega t-i\delta}\right)^2\right\}\\&=\sum_k N_k\frac{m_k}{2}\frac{(e/m)_k^2}{(\omega_k^2-\omega^2)^2}\\&\qquad\times\left\{\omega^2E_0^2\sin^2(\omega t+\delta)+\omega_k^2E_0^2\cos^2(\omega t+\delta)\right\}.\end{aligned}$$

For the time average we have the familiar result

$$\overline{|\boldsymbol{E}|^2}=\tfrac{1}{2}|\boldsymbol{E}_0|^2,$$

from which there follows

$$\overline{W}_{\text{mat}}=\frac{1}{2}\sum_k N_k\left(\frac{e^2}{m}\right)_k\frac{\omega_k^2+\omega^2}{(\omega_k^2-\omega^2)^2}\overline{|\boldsymbol{E}|^2}.$$

Thus, if we recall that $|\boldsymbol{H}|^2=n^2|\boldsymbol{E}|^2$, then there follows for the total energy density

$$W_{\text{tot}} = \frac{1}{8\pi}\overline{|\boldsymbol{E}|^2}\left\{1+n^2+4\pi\sum_k N_k\frac{e_k^2}{m_k}\frac{\omega_k^2+\omega^2}{(\omega_k^2-\omega^2)^2}\right\}.$$

According to Eq. [16.16′],

$$n^2-1 = 4\pi\sum_k N_k\frac{e_k^2}{m_k}\frac{1}{(\omega_k^2-\omega^2)}, \qquad [16.16']$$

so that the sum in the last expression is

$$\begin{aligned}4\pi\sum_k N_k\frac{e_k^2}{m_k}\frac{\omega_k^2+\omega^2}{(\omega_k^2-\omega^2)^2} &= n^2-1+4\pi\sum_k N_k\frac{e_k^2}{m_k}\frac{2\omega^2}{(\omega_k^2-\omega^2)^2}\\ &= n^2-1+\omega\frac{\mathrm{d}}{\mathrm{d}\omega}(n^2-1),\end{aligned}$$

which when substituted into the relation for $\overline{W}$ yields

$$\overline{W} = \frac{1}{8\pi}\overline{|\boldsymbol{E}|^2}\left\{2n^2+\omega\frac{\mathrm{d}}{\mathrm{d}\omega}(n^2)\right\}.$$

The Poynting vector $\boldsymbol{S}$ is

$$\boldsymbol{S} = \frac{c}{4\pi}[\boldsymbol{E}\times\boldsymbol{H}],$$

so that

$$|\boldsymbol{S}| = \frac{c}{4\pi}n|\boldsymbol{E}|^2,$$

and, furthermore,

$$\frac{c\overline{W}}{|\boldsymbol{S}|} = \frac{1}{2n}\left\{2n^2+\omega\frac{\mathrm{d}}{\mathrm{d}\omega}(n^2)\right\} = \frac{\mathrm{d}}{\mathrm{d}\omega}(n\omega).$$

On the other hand, we have $\boldsymbol{U} = \partial\omega/\partial\boldsymbol{k}$ for the group velocity. In an isotropic medium $U = \mathrm{d}\omega/\mathrm{d}k$, with $\omega = ck/n$. Thus,

$$\frac{c}{U} = c\frac{\mathrm{d}k}{\mathrm{d}\omega} = \frac{\mathrm{d}}{\mathrm{d}\omega}(n\omega),$$

from which

$$U = |\boldsymbol{S}|/\overline{W} = \mathrm{d}\omega/\mathrm{d}k. \qquad \text{Q.E.D.}$$

17. DISPERSION BY DAMPED OSCILLATORS

We start from the original form of the dispersion law

$$n^2 - 1 = 4\pi N_0 \frac{e^2}{m} \frac{1}{(\omega_0^2 - \omega^2)} \cdot \qquad [16.16]$$

This gives us a singularity at $\omega = \omega_0$. Since this, of course, cannot physically be the case, we must introduce a damping force so that the differential equation [16.10] now reads

$$m\ddot{\boldsymbol{x}} + f\boldsymbol{x} + g\dot{\boldsymbol{x}} = e\boldsymbol{E}\,, \qquad [17.1]$$

which, with

$$f = m\omega_0^2 \quad \text{and} \quad g = m\gamma\,,$$

becomes

$$\ddot{\boldsymbol{x}} + \omega_0^2 \boldsymbol{x} + \gamma\dot{\boldsymbol{x}} = \frac{e}{m}\boldsymbol{E}\,. \qquad [17.2]$$

We make the same exponential hypothesis as in Eqs. [16.11] and [16.12] so that there follows

$$\boldsymbol{x}(\omega_0^2 - \omega^2 - i\omega\gamma) = \frac{e}{m}\boldsymbol{E}\,. \qquad [17.3]$$

This, however, is exactly the same equation as [16.12′] but with ω^2 replaced by $\omega^2 + i\omega\gamma$. Thus, all conclusions remain the same if this substitution is made everywhere. Then, for example, Eq. [16.21] goes over into

$$R = \frac{4\pi}{3} A \sum_k \frac{e_k^2}{m_k} f_k \frac{1}{(\omega_k^2 - \omega^2 - i\omega\gamma)} \cdot \qquad [17.4]$$

Thus, there is no longer a pole at $\omega = \omega_k$. If the index of refraction is computed from Eq. [16.20] using [17.4], then it is seen that the index is complex. However, in accordance with the previous work, this indicates an absorption. From expression [17.4], it is seen that R^2 has an extremum (for each oscillator) when the denominator becomes minimal:

$$|\omega_0^2 - \omega^2 - i\omega\gamma|^2 = (\omega_0^2 - \omega^2)^2 + \omega^2\gamma^2$$

which is to be minimal for $\omega = \omega_m$. Then,

$$-2(\omega_0^2 - \omega_m^2)\,2\omega_m + 2\omega_m\gamma^2 = 0\,,$$

so that

$$\omega_m^2 = \omega_0^2 - \gamma^2/2\,. \qquad [17.5]$$

If there is no external field $\boldsymbol{E}$ acting on the oscillator, then it will perform free but damped oscillations. For describing these oscillations we set

$$\boldsymbol{x} = \boldsymbol{a}e^{-i\omega' t - \mu t + i\delta}\,. \qquad [17.6]$$

If Eq. [17.6] is substituted into the differential equation [17.2] with $\boldsymbol{E}=0$, there follows

$$(-i\omega' - \mu)^2 + \omega_0^2 + \gamma(-i\omega' - \mu) = 0\,.$$

We separate this into real and imaginary parts. The imaginary part is

$$2i\omega'\mu - i\omega'\gamma = 0\,,$$

so that

$$\mu = \gamma/2\,. \qquad [17.7]$$

The real part (with Eq. [17.7] substituted in) is

$$-\omega'^2 + \gamma^2/4 + \omega_0^2 - \gamma^2/2 = 0\,,$$

from which

$$\omega'^2 = \omega_0^2 - \gamma^2/4\,. \qquad [17.8]$$

Thus, the frequency of the free damped oscillation is shifted somewhat with respect to the eigenfrequency ω_0 of the oscillator.

From the theory of linear differential equations, we obtain the most general solution of Eq. [17.2] by the superposition of a particular solution, Eq. [17.3], and the general solution, Eq. [17.6], of the homogeneous equation.

The damping has the consequence that the oscillations do not become infinite at resonance.

We obtain the physical significance of the constant γ by considering the energy of the oscillation. As is well known, the energy E of an oscillator is

$$E = \frac{m}{2}\dot{x}^2 + \frac{f}{2}x^2 = \frac{m}{2}(\dot{x}^2 + \omega_0^2 x^2) .$$

The time derivative of this energy is

$$\dot{E} = m\dot{x}(\ddot{x} + \omega_0^2 x) .$$

If we assume free oscillations, then from Eq. [17.2] there follows

$$\dot{E} = -\gamma m\dot{x}^2 .$$

However, for the time average (over one period) we have

$$\bar{E} = \overline{m\dot{x}^2} ,$$

so that

$$\bar{\dot{E}} = -\gamma\bar{E} , \qquad [17.9]$$

which gives the significance of γ.

Application to the optics of metals

We use a model of a metal in which the electrons move freely in a rigid ion lattice. Thus, we must set $f=0$, $g \neq 0$ in the fundamental differential equation [17.1] which then becomes

$$m\ddot{\boldsymbol{x}} + g\dot{\boldsymbol{x}} = e\boldsymbol{E} , \qquad [17.10]$$

$$\ddot{\boldsymbol{x}} + \gamma\dot{\boldsymbol{x}} = \frac{e}{m}\boldsymbol{E} . \qquad [17.10']$$

In electrodynamics, the case where $\boldsymbol{E}$ is constant in time is considered. As the stationary solution ($\ddot{\boldsymbol{x}} = 0$), one obtains

$$\left.\begin{aligned} \dot{\boldsymbol{x}} &= \boldsymbol{v} = \frac{e}{m\gamma}\boldsymbol{E} \\ \boldsymbol{i} &= Ne\boldsymbol{v} = \sigma_0 \boldsymbol{E} \\ \sigma_0 &= \frac{Ne^2}{m\gamma} \end{aligned}\right\} . \qquad [17.11]$$

In optics, $\ddot{\boldsymbol{x}}$ must, of course, be considered too. The solution of Eq. [17.10′] is obtained in the following way. We set $\omega_0 = 0$ in all previous equations, and neglect the corrections between $\boldsymbol{E}$ and $\boldsymbol{E}'$. Then, if in addition we replace ω^2 by $\omega^2 + i\omega\gamma$ following the general scheme of this paragraph, there follows from Eq. [16.16′]

$$n'^2 - 1 = 4\pi N \frac{e^2}{m} \frac{1}{(-\omega^2 - i\omega\gamma)} . \qquad [17.12]$$

The index of refraction n' is complex, as is required by the optics of metals. For metals the optical constants were defined as follows (see Eq. [10.2]):

$$n'^2 = \varepsilon + i\frac{4\pi\sigma}{\omega} .$$

By comparing this with Eq. [17.12], there follows

$$\left.\begin{aligned} \varepsilon &= 1 - 4\pi N \frac{e^2}{m} \frac{1}{(\omega^2 + \gamma^2)} \\ \sigma &= N \frac{e^2}{m} \frac{\gamma}{(\omega^2 + \gamma^2)} \\ &= \sigma_0 \frac{\gamma^2}{(\omega^2 + \gamma^2)} \end{aligned}\right\} . \qquad [17.13]$$

The quantity γ may be frequency dependent. It can be determined theoretically only with a quantum-mechanical model.

For free electrons the damping is, in part, produced by

collisions with fixed ions; this damping decreases with decreasing density. However, even for ideal gases there is the so-called radiation damping, which is directly related to the fact that an oscillator radiates energy.[1]

18. SCATTERING OF LIGHT

a. Fundamentals from electrodynamics[1]

We start from Maxwell's equations, Eqs. [16.5], and the additional relation for polarization, Eq. [16.6]. We consider all terms to be oscillatory with the time factor $e^{-i\omega t}$ (that is, $\partial/\partial t \sim -i\omega$), and thereby obtain (for vacuum)

$$\left.\begin{aligned} \operatorname{curl} \boldsymbol{E} &= i\frac{\omega}{c}\boldsymbol{H} \\ \operatorname{curl} \boldsymbol{H} + i\omega\frac{1}{c}\boldsymbol{E} &= -\frac{4\pi i\omega}{c}\boldsymbol{P} \end{aligned}\right\}. \qquad [18.1]$$

In terms of scalar and vector potentials we have

$$\left.\begin{aligned} &\boldsymbol{H} = \operatorname{curl} \boldsymbol{A}, \qquad \boldsymbol{E} = i\frac{\omega}{c}\boldsymbol{A} - \operatorname{grad} \Phi \\ &\operatorname{div} \boldsymbol{A} - i\frac{\omega}{c}\Phi = 0 \end{aligned}\right\}. \qquad [18.2]$$

From these, there follow the second-order equations

$$\left.\begin{aligned} \nabla^2\Phi + \frac{\omega^2}{c^2}\Phi &= 4\pi \operatorname{div} \boldsymbol{P} \\ \nabla^2\boldsymbol{A} + \frac{\omega^2}{c^2}\boldsymbol{A} &= 4\pi i\frac{\omega}{c}\boldsymbol{P} \end{aligned}\right\}. \qquad [18.3]$$

The corresponding formulas for a medium with refractive index n are obtained if c is replaced by c/n and $\boldsymbol{H}$ by $\boldsymbol{H}/n$ (then Maxwell's equations remain correct). Then, from

[1] See W. PAULI, *Lectures in Physics: Electrodynamics* (M.I.T. Press, Cambridge, Mass., 1972).

Eq. [18.1] we have

$$\left.\begin{aligned} \operatorname{curl} \boldsymbol{E} &= i\frac{\omega}{c}\boldsymbol{H} \\ \operatorname{curl} \boldsymbol{H} + \frac{i\omega}{c} n^2 \boldsymbol{E} &= -\frac{4\pi}{c} i\omega n^2 \boldsymbol{P} \end{aligned}\right\}. \qquad [18.1']$$

The vector potential $\boldsymbol{A}$ should be $\boldsymbol{A}/n$ in the medium. That is, we demand that $\boldsymbol{H} = \operatorname{curl} \boldsymbol{A}$ hold in the medium as well. This, however, is arbitrary; it could be done differently. Thus, from Eq. [18.2],

$$\left.\begin{aligned} &\boldsymbol{H} = \operatorname{curl} \boldsymbol{A}, \qquad \boldsymbol{E} = i\frac{\omega}{c}\boldsymbol{A} - \operatorname{grad} \Phi \\ &\operatorname{div} \boldsymbol{A} - \frac{i\omega}{c} n^2 \Phi = 0 \end{aligned}\right\}. \qquad [18.2']$$

If we introduce the abbreviation $k = \omega n/c$, then we can also write Eq. [18.2′] as

$$\left.\begin{aligned} &\boldsymbol{H} = \operatorname{curl} \boldsymbol{A}, \qquad \boldsymbol{E} = i\frac{k}{n}\boldsymbol{A} - \operatorname{grad} \Phi \\ &\operatorname{div} \boldsymbol{A} - ikn\Phi = 0 \end{aligned}\right\}. \qquad [18.2'']$$

Furthermore, Eq. [18.3] becomes

$$\left.\begin{aligned} &\nabla^2 \Phi + k^2 \Phi = 4\pi \operatorname{div} \boldsymbol{P} \\ &\nabla^2 \boldsymbol{A} + k^2 \boldsymbol{A} = 4\pi ikn\boldsymbol{P} = 4\pi \frac{ik}{n} \boldsymbol{P}_1 \end{aligned}\right\}, \qquad [18.3']$$

where we have set

$$\boldsymbol{P}_1 = n^2 \boldsymbol{P}. \qquad [18.4]$$

If we introduce the Hertz vector $\boldsymbol{Z}$, then we have

$$\boldsymbol{A} = -ikn\boldsymbol{Z}, \qquad \Phi = -\operatorname{div} \boldsymbol{Z}. \qquad [18.5]$$

Furthermore,

$$\varrho_P = -\operatorname{div} \boldsymbol{P} \qquad [18.6]$$

(polarization charge density), and

$$\nabla^2 \boldsymbol{Z} + k^2 \boldsymbol{Z} = -4\pi \boldsymbol{P}. \tag{18.7}$$

The field intensities expressed in terms of $\boldsymbol{Z}$ are

$$\left.\begin{aligned} \boldsymbol{H} &= -ikn \operatorname{curl} \boldsymbol{Z} \\ \boldsymbol{E} &= k^2 \boldsymbol{Z} + \operatorname{grad} \operatorname{div} \boldsymbol{Z} \end{aligned}\right\}. \tag{18.8}$$

The solution of the differential equation [18.7] is

$$\boldsymbol{Z}_P = \int \boldsymbol{P}_Q \frac{e^{ikr}}{r} \, \mathrm{d}V_Q. \tag{18.9}$$

In the special case that the wavelength λ is small compared with the linear dimensions of the region in which $\boldsymbol{P}$ is appreciably different from zero, the familiar Hertz oscillator is obtained:[2]

$$\left.\begin{aligned} & R \sim |\boldsymbol{x}_P - \boldsymbol{x}_Q| \\ & \boldsymbol{\Pi} = \int \boldsymbol{P}_Q \, \mathrm{d}V_Q, \qquad \boldsymbol{\Pi}_1 = n^2 \boldsymbol{\Pi} \\ & \boldsymbol{Z}_P = \boldsymbol{\Pi} \frac{e^{ikR}}{R} \end{aligned}\right\}, \tag{18.10}$$

where P is the field point, Q the source point, and $\mathrm{d}V$ a volume element. With

$$\boldsymbol{n} = \frac{(\boldsymbol{x}_P - \boldsymbol{x}_O)}{R},$$

where O is a suitably chosen point in the source region, we have

$$\left.\begin{aligned} \boldsymbol{H} &= n \left(k^2 + \frac{ik}{R}\right) \frac{e^{ikR}}{R} (\boldsymbol{n} \times \boldsymbol{\Pi}), \\ \boldsymbol{E} &= \frac{e^{ikR}}{R} \left\{ \boldsymbol{\Pi} \left(k^2 + \frac{ik}{R} - \frac{1}{R^2}\right) \right. \\ & \qquad \left. + \boldsymbol{n}(\boldsymbol{n} \cdot \boldsymbol{\Pi}) \left(-k^2 - \frac{3ik}{R} + \frac{3}{R^2}\right) \right\} \end{aligned}\right\}. \tag{18.11}$$

[2] See PAULI, *Electrodynamics.*

The rate of energy radiation by a Hertz oscillator is

$$-\frac{\mathrm{d}E}{\mathrm{d}t}=\frac{2}{3}cnk^4|\boldsymbol{\Pi}|^2=\frac{2}{3}cn\left(\frac{\omega}{c}\right)^4|\boldsymbol{\Pi}_1|^2. \qquad [18.12]$$

b. Lorentz's theory

In his theory of scattering, Lorentz considered a homogeneous medium. In accordance with statistical mechanics, density fluctuations occur which cause scattering. The entire medium is divided into volume elements which are small compared with a cube whose edge is a wavelength long, but which, nevertheless, contain many molecules of the medium. Because of this assumption, this theory fails for x rays.

If N_V is the number of molecules in a volume element V and λ is the wavelength, then

$$N_V \gg 1 \qquad \text{and} \qquad V \ll \lambda^3 .$$

Statistically speaking, we have

$$\overline{\Delta N_V \Delta N_{V'}}=0 \qquad \text{for} \qquad V \neq V' .$$

Consequently, the light intensities originating from different volume elements may be added.

If we denote an average refractive index by $\bar{n}$, then when considering the fluctuations Eq. [18.1′] becomes

$$\operatorname{curl}\boldsymbol{H}=\{\bar{n}^2+\Delta(n^2)\}\left(-\frac{i\omega}{c}\boldsymbol{E}\right), \qquad [18.13]$$

where we have assumed that, on the average, the medium is not polarized. However, the part of $\boldsymbol{E}$ due to $\Delta(n^2)$ can be written as a polarization; indeed, upon comparison with Eq. [18.1′] this is

$$\boldsymbol{P}=\frac{1}{4\pi}\frac{\Delta(n^2)}{\bar{n}^2}\boldsymbol{E}, \qquad \boldsymbol{P}_1=\frac{1}{4\pi}\Delta(n^2)\boldsymbol{E}. \qquad [18.14]$$

Then, however, in accordance with the formula of Eq. [18.12], the volume V radiates energy:

$$-\frac{\mathrm{d}E}{\mathrm{d}t}=\frac{2cn}{3}\left(\frac{\omega}{c}\right)^4\frac{1}{16\pi^2}V^2|\boldsymbol{E}|^2\overline{[\Delta(n^2)]^2}\,.$$

This is because $\boldsymbol{\Pi}=\boldsymbol{P}V$. If the expression for the intensity, that is,

$$I=\frac{c}{4\pi}n|\boldsymbol{E}|^2,$$

is introduced into the last formula, then there follows

$$-\frac{\mathrm{d}E}{\mathrm{d}t}=\frac{1}{6\pi}\frac{\omega^4V^2}{c^4}I\,\overline{(\Delta n^2)^2}\,. \qquad [18.15]$$

We assume that the refractive index of the scattering medium is only slightly different from 1; then we can neglect the correction between $\boldsymbol{E}'$ and $\boldsymbol{E}$ and can use Eq. [16.16]. If we collect the nonfluctuating terms into the constant α, then we have

$$\overline{n}^2-1=\alpha N_V/V, \qquad [18.16]$$

$$\Delta(n^2)=\alpha\frac{\Delta N_V}{V},$$

$$\overline{(\Delta(n^2))^2}=\frac{\alpha^2}{V^2}\overline{(\Delta N_V)^2}\,, \qquad [18.17]$$

and, substituting into Eq. [18.15],

$$-\frac{\mathrm{d}E}{\mathrm{d}t}=I\frac{1}{6\pi}\frac{\omega^4}{c^4}\alpha^2\overline{(\Delta N_V)^2}\,. \qquad [18.18]$$

Equation [18.18] gives us the energy scattered per unit time from the volume V. If we consider a light ray, then its intensity loss along an interval $\mathrm{d}x$ must certainly be proportional to the energy scattered in that interval. Therefore,

$$-\frac{\mathrm{d}E}{\mathrm{d}t}=-\frac{\mathrm{d}I}{\mathrm{d}x}V, \qquad [18.18']$$

as can easily be seen. With this, from Eq. [18.18],

$$-\frac{\mathrm{d}I}{\mathrm{d}x}=\frac{1}{6\pi}\frac{\omega^4}{c^4}\alpha^2\frac{I}{V}\overline{(\Delta N_V)^2}. \qquad [18.19]$$

It is customary to introduce the *scattering coefficient* s and to substitute the expression for α given by [18.16] into Eq. [18.19]. From this there follows

$$-\frac{\mathrm{d}I}{\mathrm{d}x}=sI, \qquad [18.20]$$

with

$$s=\frac{8\pi^3}{3}(\overline{n}^2-1)^2\frac{1}{\overline{N}^2}\frac{1}{\lambda_0^4}\frac{\overline{(\Delta N_V)^2}}{V},$$

where we have used the abbreviation $\overline{N}\equiv N_V/V$.

From statistical mechanics (the theory of fluctuations) we have the formula[3]

$$\overline{(\Delta N_V)^2}=\frac{kT}{(-\partial p/\partial V)_{N_V,T}(V/N_V)^2}, \qquad [18.21]$$

or

$$\frac{\overline{(\Delta N_V)^2}}{N_V}=\frac{\overline{N}kT}{-V(\partial p/\partial V)_{N_V,T}}.$$

If this is substituted into the equation for the scattering coefficient, there follows

$$s=\frac{8\pi^3}{3}\frac{(\overline{n}^2-1)^2}{\lambda_0^4}\left(\frac{kT}{-V(\partial p/\partial V)_{N_V,T}}\right). \qquad [18.22]$$

For ideal gases

$$-V\left(\frac{\partial p}{\partial V}\right)_{N_V,T}=\overline{N}kT,$$

so that for these the scattering coefficient is

$$s=\frac{8\pi^3}{3}\frac{(\overline{n}^2-1)^2}{\overline{N}\lambda_0^4}. \qquad [18.23]$$

Formula [18.23] permits Avogadro's number to be deter-

[3] See W. Pauli, *Lectures in Physics: Statistical Mechanics* (M.I.T. Press Cambridge, Mass., 1972).

mined. Furthermore, we see that short waves are scattered more strongly than long waves; the blue color of the sky is a result of this.

c. Scattering of x rays

We assume that the wavelength λ is small compared with the dimensions of an atom. Then, we have incoherent scattering from the electrons. Let S be the energy scattered per unit time, per unit volume. Then we have (radiation from moving charges[4])

$$S = \frac{2}{3} \frac{e^2}{c^3} \dot{v}^2 NZ, \qquad [18.24]$$

in which N represents the number of atoms per unit volume, Z the number of electrons per atom, and v the velocity of the electrons.

Now, the frequency of light is much greater than the eigenfrequencies of the electrons in the atom. Thus, we set ω_0 (and ω_k) equal to zero. This results in assuming (almost) free electrons. For such electrons,

$$m\dot{\boldsymbol{v}} = e\boldsymbol{E}. \qquad [18.25]$$

We substitute Eq. [18.25] and the relation for the incident intensity, $I = (c/4\pi) \boldsymbol{E}^2$, into Eq. [18.24]:

$$S = \frac{8\pi}{3} \frac{e^4}{m^2 c^4} INZ. \qquad [18.26]$$

The scattering coefficient is given by $s = S/I$ (this follows simply from the definition in Eq. [18.20] together with [18.18′]). Therefore,

$$s = \frac{8\pi}{3} \left(\frac{e^2}{mc^2}\right)^2 NZ. \qquad [18.27]$$

Now, N can be expressed in terms of the density ϱ, the

[4] See PAULI, *Electrodynamics*.

molecular weight M, and Avogadro's number A:

$$N = \varrho A / M .$$

If this is substituted into Eq. [18.27] and if one then divides by the density, then the *mass absorption coefficient* σ is obtained:

$$\sigma \equiv \frac{s}{\varrho} = \frac{8\pi}{3}\left(\frac{e^2}{mc^2}\right)^2 \frac{A}{M} Z = 0.4 \frac{Z}{M} . \qquad [18.28]$$

Equation [18.28] was established by Thomson. The number of electrons in an atom Z can be determined with it.

The dispersion formula for x rays is obtained by setting ω_0 equal to zero in Eq. [16.16]:

$$n^2 - 1 = -4\pi N Z \frac{e^2}{m\omega^2} .$$

This shows the possibility that x rays may be totally reflected.

Refinement. We consider a polarized wave, whose $\boldsymbol{E}$ vector is parallel to the z axis, incident on an atom. As before, the electron is considered to be "almost free"; however, we now want to consider the fact that not all of the Z electrons of the atom that scatter the radiation are located at O, the position of the nucleus. The effect of the incident wave is to produce a periodic acceleration of the electron, in accordance with Eq. [18.25]. However, when accelerated periodically, the electron produces an outgoing (scattered) wave whose field intensity $\boldsymbol{E}_P$ at a distance r (from the electron) is calculated in electromagnetic theory:

$$\boldsymbol{E}_P = \frac{e}{c^2} \cdot \frac{e^{ikr}}{r} \dot{\boldsymbol{v}}_\perp . \qquad [18.29]$$

The notation $\perp$ is to be understood as denoting the component of the acceleration $\dot{\boldsymbol{v}}$ which is perpendicular to the direction of propagation under consideration. Consequently, the magnitude of the field intensity $\boldsymbol{E}_P$ at the point P (a distance r from the electron, making an angle θ with x axis) is

$$E_P = \frac{e^2}{mc^2} \cdot E_e(\boldsymbol{x}) \frac{e^{ikr}}{r} \cos\theta .$$

Here, $E_e(\boldsymbol{x})$ is the magnitude of the incident field intensity at the point $\boldsymbol{x}$ and $\boldsymbol{E}_P$, perpendicular to the direction of scattering, lies in the plane containing this direction and the direction of oscillation of the electron. If we assume that $r \gg |\boldsymbol{x}|$, then the distance $R = |\overrightarrow{PO}|$ becomes

$$R = r + (\boldsymbol{n} \cdot \boldsymbol{x}) + \dots ,$$

where $\boldsymbol{n}$ denotes a unit vector in the direction of the scattering. Hence, we have (as an approximation)

$$E_P = \frac{e^2}{mc^2} \cdot E_e(\boldsymbol{x})\, e^{-i\boldsymbol{k}\cdot\boldsymbol{x}} \frac{e^{ikR}}{R} \cos\theta .$$

If we write the wave vector of the incident wave as

$$\boldsymbol{k}_e = k\boldsymbol{n}_e ,$$

then, in our approximation,

$$\boldsymbol{E}_e(\boldsymbol{x}) = \boldsymbol{E}_e^0 e^{i\boldsymbol{k}_e\cdot\boldsymbol{x}} ,$$

where $\boldsymbol{E}_e^0$ denotes the field intensity of the incident wave at the origin. From this there follows

$$E_P = \frac{e^2}{mc^2} \cdot E_e^0 e^{i(\boldsymbol{k}_e - \boldsymbol{k})\cdot\boldsymbol{x}} \frac{e^{ikR}\cos\theta}{R} . \qquad [18.30]$$

If we assume that there are several electrons present in an atom, then we must write

$$E_P = \frac{e^2}{mc^2} E_e^0 \sum_n e^{i(\boldsymbol{k}_e - \boldsymbol{k})\cdot\boldsymbol{x}_n} \frac{e^{ikR}}{R} \cos\theta , \qquad [18.30']$$

where the sum is taken over the Z electrons. Of course,

E_e^0 is periodic in time with the factor $e^{-i\omega t}$:

$$E_e^0 = F e^{-i(\omega t + \delta)}.$$

The intensity of the incident wave is then

$$I_e = \frac{c}{4\pi} \cdot F^2.$$

The intensity scattered from the atom into the solid angle $\mathrm{d}\Omega$ is

$$\mathrm{d}I = I_e \left(\frac{e^2}{mc^2}\right)^2 \cos^2\theta |K|^2 \mathrm{d}\Omega. \qquad [18.31]$$

The factor

$$K = \sum_n e^{i(\boldsymbol{k}_e - \boldsymbol{k}) \cdot \boldsymbol{x}_n} \qquad [18.32]$$

is called the structure factor.

If we have light which is polarized at an angle with respect to the plane defined by the incident and the scattered directions, then we must decompose its electric field vector into components parallel and perpendicular to this plane. The component parallel to the plane is scattered according to Eqs. [18.31] and [18.32]; the perpendicular component is scattered according to the same formulas evaluated at $\theta = 0$. From this we see that the entire calculation for this component is the same as before, except for the single difference that in Eq. [18.29], $\dot{\boldsymbol{v}}_\perp$ is the same as $\dot{\boldsymbol{v}}$. That is, $\cos\theta$ is to be replaced by unity.

The square of the magnitude of the structure factor can be calculated:

$$|K|^2 = K^* K = \sum_{n,m} e^{i(\boldsymbol{k}_e - \boldsymbol{k}) \cdot \boldsymbol{x}_n} e^{-i(\boldsymbol{k}_e - \boldsymbol{k}) \boldsymbol{x}_m}$$

$$= Z + 2 \sum_{n<m} \cos\left[k(\boldsymbol{n}_e - \boldsymbol{n}) \cdot (\boldsymbol{x}_n - \boldsymbol{x}_m)\right], \qquad [18.33]$$

where Z represents the number of electrons in the atom.

Now, if

$$k(\boldsymbol{n}_e - \boldsymbol{n}) \cdot (\boldsymbol{x}_n - \boldsymbol{x}_m) \gg 1,$$

or

$$(\boldsymbol{n}_e - \boldsymbol{n}) \cdot (\boldsymbol{x}_n - \boldsymbol{x}_m) \gg \frac{\lambda}{2\pi}, \qquad [18.34]$$

then all possible phases occur as arguments of the cosine and, consequently, the sum is small. Thus,

$$|K|^2 \sim Z. \qquad [18.35]$$

If the incident light is unpolarized, then all directions of $\boldsymbol{E}_e^0$ are equally probable and the incident intensity is divided equally among the two states of polarization. Thus, Eq. [18.31] can be integrated. There follows

$$I = \frac{1}{2} I_e \left[\int \left(\frac{e^2}{mc^2} \right)^2 \cos^2\theta \, Z \, \mathrm{d}\Omega + \int \left(\frac{e^2}{mc^2} \right)^2 Z \, \mathrm{d}\Omega \right]$$

$$= \frac{1}{2} I_e Z \left(\frac{e^2}{mc^2} \right)^2 \left[\int \cos^2\theta \sin\theta \, \mathrm{d}\theta + \int \sin\theta \, \mathrm{d}\theta \right] 2\pi,$$

so that

$$I = \frac{8\pi}{3} \left(\frac{e^2}{mc^2} \right)^2 Z I_e,$$

and this is the previous formula of Eq. [18.27]. The inequality [18.34] is always unsatisfied within a certain cone which, however, becomes smaller as λ decreases.

Remarks. According to the classical theory as well as to the quantum theory, the electrons in an atom are not fixed but, instead, have a certain probability of being at a given position. For a particular electron configuration, let this probability be

$$W(\boldsymbol{x}_1, \ldots, \boldsymbol{x}_Z) \, \mathrm{d}V_1 \ldots \mathrm{d}V_Z,$$

with

$$\int \ldots \int W \, \mathrm{d}V_1 \ldots \mathrm{d}V_Z = 1.$$

Then. the structure factor is to be taken as a statistical

average:

$$\overline{|K|^2} = \int \cdots \int W \left| \sum_n e^{i(\boldsymbol{k}_e - \boldsymbol{k}) \cdot \boldsymbol{x}_n} \right|^2 \mathrm{d}V_1 \ldots \mathrm{d}V_Z . \qquad [18.36]$$

If W separates into a product,

$$\left. \begin{aligned} W(\boldsymbol{x}_1, \ldots, \boldsymbol{x}_Z) &= w_1(\boldsymbol{x}_1) w_2(\boldsymbol{x}_2) \ldots w_Z(\boldsymbol{x}_Z) \\ \text{with} \qquad \int w_i(\boldsymbol{x}_i)\, \mathrm{d}V_i &= 1 , \end{aligned} \right\} \qquad [18.37]$$

then expression [18.36] can be simplified. If we define the abbreviation

$$F_n = \int w_n(\boldsymbol{x}_n) e^{i(\boldsymbol{k}_e - \boldsymbol{k}) \cdot \boldsymbol{x}_n} \mathrm{d}V_n , \qquad [18.38]$$

then

$$\overline{|K|^2} = \sum_{\substack{n \quad m \\ n \neq m}} \sum F_n F_m^* + Z = Z + \sum_{n<m} (F_n F_m^* + F_n^* F_m) . \qquad [18.39]$$

If, in addition, w_n depends only upon r_n,

$$w_n(\boldsymbol{x}_n) = w_n(r_n) ,$$

that is, if w_n is spherically symmetric, then Eq. [18.38] can be evaluated further. With θ representing the scattering angle,

$$|\boldsymbol{k}_e - \boldsymbol{k}| = 2k \sin \theta/2 ,$$

$$\alpha = \sphericalangle \{(\boldsymbol{k}_e - \boldsymbol{k}), \boldsymbol{x}\} ,$$

$$\cos \alpha = u ,$$

so that:

$$\mathrm{d}V = r^2 \, \mathrm{d}r 2\pi \, \mathrm{d}u ,$$

$$F_n = \int w_n(r_n) \frac{\sin(2kr \sin \theta/2)}{kr \sin \theta/2} 2\pi r^2 \, \mathrm{d}r . \qquad [18.38']$$

Thus, for spherically symmetric distributions, F_n is only a function of $(4\pi/\lambda)\sin\theta/2$:

$$F_n = F_n\left(\frac{4\pi}{\lambda}\sin\theta/2\right). \qquad [18.38'']$$

Here, if λ again approaches zero (more exactly, $(\sin\theta/2)/\lambda \rightarrow \infty$), then F_n approaches zero and, as a result, $\overline{|K|^2}$ approaches Z, which is the same result as before.

In the above considerations we neglected the fact that (according to classical theory) a Doppler effect occurs during scattering. This effect shifts the frequency of the scattered radiation somewhat with respect to that of the incident radiation.

19. OPTICAL ACTIVITY

a. Phenomenological theory

The optical activity of a substance consists of the rotation of the plane of oscillation of polarized light upon passing through the substance. This effect can be described by saying that the refractive indices for waves of opposite circular polarization are different. (The superposition of two waves of opposite circular polarization of equal amplitude and frequency clearly results in a linearly polarized wave.) The angle χ through which the plane of oscillation is rotated is obviously

$$\chi = \frac{\omega}{c}\,\frac{(n_+ - n_-)}{2}\,l\,, \qquad [19.1]$$

where l is the path traveled in the optically active substance, and n_+ and n_- are the refractive indices for right-hand and left-hand circularly polarized light.

In order for optical activity to occur, it is of course necessary that the optical equations no longer be invariant under reflection. This results from the fact that certain molecules are not mirror-symmetric (for example, sugar solution, tourmaline crystal), so that the eigenfrequencies for right-hand and left-hand circularly polarized light are different.

Ordinary (polar) vectors are invariant under reflections, whereas axial vectors, such as the vector products of ordinary vectors and the magnetic field vector, are not. As shown in *Electrodynamics*,[5] these vectors are actually skew-symmetric tensors which change sign upon reflection.

We take the following (Maxwell) equations from electrodynamics:

$$\left.\begin{aligned} \operatorname{curl} \boldsymbol{E} &= -\frac{1}{c}\dot{\boldsymbol{B}} \\ \operatorname{curl} \boldsymbol{B} &= \frac{1}{c}\dot{\boldsymbol{E}} + \frac{4\pi}{c}\boldsymbol{I}_M \end{aligned}\right|, \qquad [19.2]$$

$$\left.\begin{aligned} \boldsymbol{H} &= \boldsymbol{B} - 4\pi\boldsymbol{M} \\ \boldsymbol{I}_M &= \dot{\boldsymbol{P}} + c \operatorname{curl} \boldsymbol{M} \end{aligned}\right\}. \qquad [19.3]$$

Furthermore, we assume

$$\left.\begin{aligned} \boldsymbol{P} &= \frac{\varepsilon - 1}{4\pi}\boldsymbol{E} - \alpha_1 \operatorname{curl} \boldsymbol{E} \\ \boldsymbol{M} &= -\alpha_2 \frac{1}{c}\dot{\boldsymbol{E}} \\ &\alpha_1 + \alpha_2 = \alpha \end{aligned}\right|. \qquad [19.4]$$

There then follows

$$\left.\begin{aligned} \boldsymbol{I}_M &= \frac{\varepsilon - 1}{4\pi}\dot{\boldsymbol{E}} - \alpha \operatorname{curl} \dot{\boldsymbol{E}} \\ \boldsymbol{D} &= \varepsilon\boldsymbol{E} - 4\pi\alpha \operatorname{curl} \boldsymbol{E} \end{aligned}\right|, \qquad [19.5]$$

where $\boldsymbol{D}$ satisfies the relation

$$\operatorname{curl} \boldsymbol{B} = \frac{1}{c}\dot{\boldsymbol{D}}. \qquad [19.6]$$

With these equations, the following results obtain for a

[5] Ibid.

plane wave ($\nabla \sim i\boldsymbol{k}$, $\partial/\partial t \sim -i\omega$):

$$\boldsymbol{E} = \boldsymbol{E}_0 e^{i(k\boldsymbol{n}\cdot\boldsymbol{x}-\omega t)},$$

with

$$\boldsymbol{k} = \frac{\omega}{c} n\boldsymbol{n}.$$

With the abbreviation

$$-4\pi\alpha\frac{\omega}{c} n = f, \qquad [19.7]$$

there follows

$$\boldsymbol{D} = \varepsilon\boldsymbol{E} + if\boldsymbol{n}\times\boldsymbol{E}. \qquad [19.8]$$

Now, as is well known, the Maxwell equations yield

$$\text{curl curl}\,\boldsymbol{E} = -\frac{1}{c^2}\ddot{\boldsymbol{D}}$$

and so, for a plane wave (where $\boldsymbol{E}\cdot\boldsymbol{n} = 0$), there follows

$$-n^2\boldsymbol{n}\times(\boldsymbol{n}\times\boldsymbol{E}) = -n^2[(\boldsymbol{n}\cdot\boldsymbol{E})\boldsymbol{n} - \boldsymbol{E}] = \boldsymbol{D},$$

so that with Eq. [19.8],

$$n^2\boldsymbol{E} = \varepsilon\boldsymbol{E} + if\boldsymbol{n}\times\boldsymbol{E}. \qquad [19.9]$$

We now locate $\boldsymbol{n}$ along the z axis $(\boldsymbol{n} = \{0, 0, 1\})$ and $\boldsymbol{E}$ perpendicular to it $(\boldsymbol{E} = \{E_x, E_y, 0\})$. Equation [19.9] yields

$$\left.\begin{aligned} n^2E_x &= \varepsilon E_x - ifE_y \\ n^2E_y &= \varepsilon E_y + ifE_x \end{aligned}\right\}. \qquad [19.10]$$

In the usual way, we introduce E_{I} and E_{II} in place of E_x and E_y:

$$\left.\begin{aligned} E_{\mathrm{I}} &= E_x + iE_y \\ E_{\mathrm{II}} &= E_x - iE_y \end{aligned}\right\}. \qquad [19.11]$$

Then,

$$\left.\begin{aligned} \{n^2 - \varepsilon + f\}E_{\mathrm{I}} &= 0 \\ \{n^2 - \varepsilon - f\}E_{\mathrm{II}} &= 0 \end{aligned}\right\}, \qquad [19.10']$$

or, written explicitly,

$$\left.\begin{aligned}\left(n^2 - \varepsilon - 4\pi\alpha\frac{\omega}{c}\,n\right)E_{\mathrm{I}} &= 0\\ \left(n^2 - \varepsilon + 4\pi\alpha\frac{\omega}{c}\,n\right)E_{\mathrm{II}} &= 0\end{aligned}\right\}. \qquad [19.10'']$$

This new system has two solutions:

$$(1)\quad E_{\mathrm{II}} = 0; \qquad E_x = iE_y\,, \qquad n = n_{\mathrm{I}}\,, \qquad [19.12]$$

$$n_{\mathrm{I}}^2 - \varepsilon - 4\pi\alpha\,\frac{\omega}{c}\,n = 0\,,$$

so that (with α sufficiently small)

$$n_{\mathrm{I}} \simeq \sqrt{\varepsilon} + 2\pi\alpha\frac{\omega}{c}\,. \qquad [19.12']$$

This produces a left-hand circularly polarized wave (in accordance with Eq. [19.12]). The corresponding index of refraction is given by Eq. [19.12′].

$$(2)\quad E_{\mathrm{I}} = 0; \qquad E_x = -\,iE_y\,, \qquad n = n_{\mathrm{II}}\,.$$

From a calculation analogous to the preceding one, there follows

$$n_{\mathrm{II}} \simeq \sqrt{\varepsilon} - 2\pi\alpha\,\frac{\omega}{c}\,.$$

Thus, from Eq. [19.1], the rotation of the plane of oscillation becomes

$$\chi = -\,2\pi\alpha\left(\frac{\omega}{c}\right)^2 l\,. \qquad [19.1']$$

In our theory, conservation of energy follows correctly if we make the following assumptions:

$$S = \frac{c}{4\pi}\,\boldsymbol{E}\times\left(\boldsymbol{B} + \frac{2\pi}{c}\,\alpha\dot{\boldsymbol{E}}\right),$$

$$W = \frac{1}{8\pi}\,\{\varepsilon\boldsymbol{E}^2 + \boldsymbol{B}^2 - 4\pi\alpha\boldsymbol{E}\cdot\operatorname{curl}\boldsymbol{E}\}\,.$$

The essential feature of this theory is that in the fundamental assumption, Eq. [19.4], we establish a relationship between a proper vector and one which has the transformation character of a vector product.

In the case of optically active *crystals*, the relation between $\boldsymbol{D}$ and $\boldsymbol{E}$ must be made more general:

$$D_i = \sum_k \eta_{ik} E_k . \qquad [19.8a]$$

Conservation of energy requires that the tensor η_{ik} be hermitian,

$$\eta_{ki} = \eta_{ik}^* ,$$

so that it can be decomposed in the following manner:

$$\eta_{ik} = \varepsilon_{ik} + i\varrho_{ik} ,$$

with $\varepsilon_{ik} = \varepsilon_{ki}$, and $\varrho_{ik} = -\varrho_{ki}$. Consequently, we can write further

$$\varrho_{ik} = \sum_l f_{ik,l} n_l , \qquad [19.8b]$$

with

$$f_{ki,l} = - f_{ik,l} .$$

Equation [19.8a] can also be written with the help of a vector product. Let $\boldsymbol{\rho}$ be the vector associated with the skew-symmetric tensor ϱ_{ik}. That is, let

$$\varrho_i = \sum_{k<l} \varepsilon_{ikl} \varrho_{kl} ,$$

where ε_{ikl} denotes the familiar permutation tensor:

$$\varepsilon_{ikl} = \begin{cases} +1 \text{ if } i, k, l \text{ is an even permutation of } 1, 2, 3; \\ -1 \text{ if } i, k, l \text{ is an odd permutation of } 1, 2, 3; \\ \;\;\,0 \text{ if any two of the indices are equal.} \end{cases}$$

Then, Eq. [19.8a] goes over into

$$D_j = \sum_k \varepsilon_{jk} E_k - i(\boldsymbol{\rho} \times \boldsymbol{E})_j$$

and the formula corresponding to Eq. [19.8b] reads

$$\varrho_i = \sum_k f_{ik} n_k \, .$$

The equations for isotropic media are re-obtained from these two formulas if we set

$$f_{ik} = f\delta_{ik} \, , \qquad f_{ik,l} = f\varepsilon_{ikl} \, , \qquad \text{and} \qquad \varepsilon_{ik} = \varepsilon\delta_{ik} \, .$$

b. Kuhn's model [A-7]

As a model of an optically active molecule we take two coupled oscillators which, in what follows, are to be arranged in a mirror noninvariant way (see Fig. 19.1). To

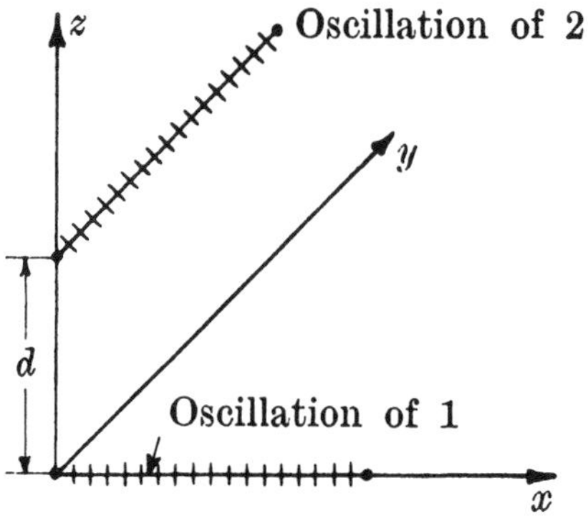

Figure 19.1

simplify the calculation, we assume equal masses for the two oscillators. Then, taking the coupling into account, the potential energy of the system becomes

$$E_{\text{pot}} = \frac{m}{2} (\omega_1^2 x_1^2 + \omega_2^2 y_2^2 + k x_1 y_2) \, .$$

For free oscillations (in the absence of an electromagnetic

field) we obtain

$$\ddot{x}_1 = -\omega_1^2 x_1 - k y_2 - (1/m)\partial E_{\text{pot}}/\partial x_1 ,$$

$$\ddot{y}_2 = -k x_1 - \omega_2^2 y_2 - (1/m)\,\partial E_{\text{pot}}/\partial y_2 .$$

With the assumption

$$x_1 = A e^{i\omega_0 t} , \qquad y_2 = B e^{i\omega_0 t} ,$$

there follows

$$\left.\begin{aligned} x_1(-\omega_0^2 + \omega_1^2) + y_2 k = 0 \\ x_1 k + y_2(-\omega_0^2 + \omega_2^2) = 0 \end{aligned}\right\} , \qquad [19.13]$$

which, if a nontrivial solution is to exist for x_1 and y_2, leads to the determinantal requirement

$$(-\omega_0^2 + \omega_1^2)(-\omega_0^2 + \omega_2^2) - k^2 = 0 . \qquad [19.13']$$

The roots ω' and ω'' will be in the neighborhood of the eigenfrequencies ω_1 and ω_2, corresponding to the uncoupled oscillators. Thus, we have the expansions

$$\left.\begin{aligned} \omega'^2 \simeq \omega_1^2 - \frac{k^2}{\omega_2^2 - \omega_1^2} \\ \omega''^2 \simeq \omega_2^2 + \frac{k^2}{\omega_2^2 - \omega_1^2} \end{aligned}\right\} . \qquad [19.14]$$

For the two eigenoscillations of the system, there follows from Eq. [19.13]

$$\left.\begin{aligned} y_2 = \frac{-k}{\omega_2^2 - \omega_1^2} x_1 \\ x_1 = \frac{+k}{\omega_2^2 - \omega_1^2} y_2 \end{aligned}\right\} . \qquad [19.15]$$

If we now assume light incident parallel to the z axis, then we have a forced oscillation. With the same assumption as before, instead of Eq. [19.13] we now obtain

$$\left.\begin{aligned}(-\omega^2+\omega_1^2)x_1+ky_2&=\frac{e}{m}E_{1x}\\ kx_1+(-\omega^2+\omega_2^2)y_2&=\frac{e}{m}E_{2y}\end{aligned}\right\}, \qquad [19.16]$$

where ω is the frequency of the light and all quantities are periodic with this frequency. This yields

$$\left.\begin{aligned}&[(-\omega^2+\omega_1^2)(-\omega^2+\omega_2^2)-k^2]x_1\\ &\qquad=\frac{e}{m}[(-\omega^2+\omega_2^2)E_{1x}-kE_{2y}]\\ &[(-\omega^2+\omega_1^2)(-\omega^2+\omega_2^2)-k^2]y_2\\ &\qquad=\frac{e}{m}[(-\omega^2+\omega_1^2)E_{2y}-kE_{1x}]\end{aligned}\right\}. \qquad [19.16']$$

On the basis of Eq. [19.13′], we can write the first square bracket in Eq. [19.16′] as

$$(\omega^2-\omega'^2)(\omega^2-\omega''^2)\,.$$

Thus, there follows for the electric dipole moment

$$\left.\begin{aligned}P_x&\equiv\sum_k ex_k=\frac{e^2/m}{(\omega^2-\omega'^2)(\omega^2-\omega''^2)}\\ &\qquad\qquad\times[(-\omega^2+\omega_2^2)E_{1x}-kE_{2y}]\\ P_y&\equiv\sum_k ey_k=\frac{e^2/m}{(\omega^2-\omega'^2)(\omega^2-\omega''^2)}\\ &\qquad\qquad\times[(-\omega^2+\omega_1^2)E_{2y}-kE_{1x}]\\ P_z&\equiv\sum_k ez_k=\text{constant}\end{aligned}\right\}. \qquad [19.17]$$

If we denote the electric field intensity of the incident wave at the center of the interval d by $\boldsymbol{E}$, then we can expand and obtain

$$E_{1x}=E_x-\frac{d}{2}\frac{\partial E_x}{\partial z}-\dots,$$

$$E_{2y}=E_y+\frac{d}{2}\frac{\partial E_y}{\partial z}+\dots.$$

Thus,

$$\left.\begin{aligned} P_x &= \frac{e^2/m}{(\omega^2-\omega'^2)(\omega^2-\omega''^2)}\left[(-\omega^2+\omega_2^2)E_x \right.\\ &\qquad \left. - kE_y - k\frac{d}{2}\frac{\partial E_y}{\partial z} - (-\omega^2+\omega_2^2)\frac{d}{2}\frac{\partial E_x}{\partial z}\right] \\ P_y &= \frac{e^2/m}{(\omega^2-\omega'^2)(\omega^2-\omega''^2)}\left[-kE_x \right.\\ &\qquad \left. + (-\omega^2+\omega_1^2)E_y + (-\omega^2+\omega_1^2)\frac{d}{2}\frac{\partial E_y}{\partial z} + k\frac{d}{2}\frac{\partial E_x}{\partial z}\right] \end{aligned}\right\}. \quad [19\ 17']$$

For anisotropic media, this cannot be further simplified. However, for isotropic media we can average over the positions of the molecules and can calculate the average value of $\boldsymbol{P}$. The average value of $\boldsymbol{P}$ for all possible orientations of the oscillators parallel to the x-y plane (denoted by $\boldsymbol{P}'$) is obtained by forming the arithmetic mean of $\boldsymbol{P}$ for the original position and $\boldsymbol{P}$ for a position rotated by 90°:

$$\boldsymbol{P}' = \frac{e^2/m}{(\omega^2-\omega'^2)(\omega^2-\omega''^2)}\left[\left(-\omega^2+\frac{\omega_1^2+\omega_2^2}{2}\right)\boldsymbol{E} - \frac{\omega_2^2-\omega_1^2}{2}\frac{d}{2}\frac{\partial \boldsymbol{E}}{\partial z} + k\frac{d}{2}\operatorname{curl}\boldsymbol{E}\right].$$

The average value (denoted by $\overline{\boldsymbol{P}}$) for the possible positions of the axes of the oscillators (coinciding with the z axis in the figure) is obtained as an arithmetic mean of $\boldsymbol{P}'$ for the two oppositely directed positions of the axis. By calculating $\boldsymbol{P}'$, making the substitution $x \to -x$, $y \to y$, $z \to -z$ and recalculating $\boldsymbol{P}'$, adding the two and taking half the sum, we get

$$\overline{\boldsymbol{P}} = \frac{e^2/m}{(\omega^2-\omega'^2)(\omega^2-\omega''^2)} \times\left[\left(-\omega^2+\frac{\omega_1^2+\omega_2^2}{2}\right)\boldsymbol{E} + \frac{kd}{2}\operatorname{curl}\boldsymbol{E}\right]. \qquad [19.18]$$

In a completely similar manner we now calculate the

magnetic dipole moment,

$$\boldsymbol{M} \equiv \frac{e}{2c} \sum_k \boldsymbol{x}_k \times \dot{\boldsymbol{x}}_k .$$

The magnetic moment must be referred to the same point as the electric moment—that is, to the center of the interval d. There then follows

$$\left.\begin{aligned} M_x &= -\frac{e}{2c}\frac{d}{2}\dot{y}_2 \\ M_y &= -\frac{e}{2c}\frac{d}{2}\dot{x}_1 \\ M_z &= 0 \end{aligned}\right\} . \tag{19.19}$$

We now again expand $\boldsymbol{E}$ about the center of the interval d and substitute into Eq. [19.19]. (See Eq. [19.17] above.) For this, we need go only up to the constant term in the Taylor expansion, because Eq. [19.19] is already linear in d and we consistently neglect higher-order terms. Thus, there follows

$$\left.\begin{aligned} M_x &= -\frac{e}{2c}\frac{d}{c}\dot{y}_2 = -\frac{d}{4c}\dot{P}_y \\ &= -\frac{d}{4c}\frac{e^2/m}{(\omega^2-\omega'^2)(\omega^2-\omega''^2)} \\ &\qquad \times[(-\omega^2+\omega_1^2)\dot{E}_y - k\dot{E}_x] \\ M_y &= -\frac{d}{4c}\frac{e^2/m}{(\omega^2-\omega'^2)(\omega^2-\omega''^2)} \\ &\qquad \times[(-\omega^2+\omega_2^2)\dot{E}_x - k\dot{E}_y] \end{aligned}\right\} . \tag{19.20}$$

As in the preceding calculations, we must average, first over the positions of the oscillators parallel to the x-y plane:

$$\left.\begin{aligned} M'_x &= -\frac{d}{4c}\frac{e^2/m}{(\omega^2-\omega'^2)(\omega^2-\omega''^2)} \\ &\qquad \times[-k\dot{E}_x + \tfrac{1}{2}(\omega_1^2-\omega_2^2)\dot{E}_y] \\ M'_y &= -\frac{d}{4c}\frac{e^2/m}{(\omega^2-\omega'^2)(\omega^2-\omega''^2)} \\ &\qquad \times[-\tfrac{1}{2}(\omega_2^2-\omega_1^2)\dot{E}_x - k\dot{E}_y] \end{aligned}\right\} , \tag{19.20'}$$

and then over the directions of the axes:

$$\overline{\boldsymbol{M}} = \frac{d}{4c} \frac{e^2/m}{(\omega^2 - \omega'^2)(\omega^2 - \omega''^2)} k\dot{\boldsymbol{E}}. \qquad [19.21]$$

Thus, we see that Kuhn's model explains the phenomenological quantities α_1 and α_2. By comparison with Eq. [19.4], there follows (for isotropic bodies)

$$\left.\begin{aligned} \alpha_1 &= \frac{e^2/m}{(\omega^2 - \omega'^2)(\omega^2 - \omega''^2)} k\frac{d}{2} \\ \alpha_2 &= -\frac{e^2/m}{(\omega^2 - \omega'^2)(\omega^2 - \omega''^2)} k\frac{d}{4} \\ \alpha &= \frac{e^2/m}{(\omega^2 - \omega'^2)(\omega^2 - \omega''^2)} k\frac{d}{4} \end{aligned}\right\}. \qquad [19.22]$$

To summarize, we want to emphasize that in Kuhn's model the coupling of the oscillators and the position dependence of the field intensities of the incident wave are essential. The decomposition of the effect into a part from $\boldsymbol{P}$ and a part from $\boldsymbol{M}$ is not essential; they can be combined by introducing the concept of the retarded electric moment.[6] The fact that the finite velocity of propagation of the force between the two oscillators has not been considered is also unessential. This is true because it can be shown that if the forces in the atom are electrical in nature, this effect is of order of magnitude d^2 and, consequently, is to be neglected. Also unessential is the fact that we had only two coupled oscillators. For the general case, see M. Born, *Optik* (Supplementary Bibliography).

20. MAGNETO-OPTICS

a. Zeeman effect

In the following, we will study the oscillations of a particle in a static magnetic field $\boldsymbol{H}$.

[6] See Krooy, *Dissertation*, Leyden (1936).

The fundamental equation is obtained by assuming the Lorentz force:

$$\ddot{\boldsymbol{x}}^2 + \omega_0^2 \boldsymbol{x} = \frac{e}{mc}\,\dot{\boldsymbol{x}} \times \boldsymbol{H}. \qquad [20.1]$$

If we take the z axis parallel to the magnetic field $\boldsymbol{H}$ so that $\boldsymbol{H} = \{0, 0, H\}$, then there follows:
1. For the z component:

$$\ddot{z} + \omega_0^2 z = 0\,,$$

$$z = Ce^{-i\omega_0 t}\,.$$

This is an oscillation parallel to the magnetic field with unchanged frequency ω_0.
2. For the x and y components:

$$\ddot{x} + \omega_0^2 x = \frac{eH}{mc}\,\dot{y}\,,$$

$$\ddot{y} + \omega_0^2 y = -\frac{eH}{mc}\,\dot{x}\,.$$

With the substitutions

$$\xi_{\mathrm{I}} = x + iy\,,$$

$$\xi_{\mathrm{II}} = \xi_{\mathrm{I}}^* = x - iy\,,$$

we get

$$\ddot{\xi}_{\mathrm{I}} + \omega_0^2 \xi_{\mathrm{I}} = -\frac{eH}{mc}\,i\dot{\xi}_{\mathrm{I}}\,,$$

$$\ddot{\xi}_{\mathrm{II}} + \omega_0^2 \xi_{\mathrm{II}} = +\frac{eH}{mc}\,i\dot{\xi}_{\mathrm{II}}\,.$$

If we assume

$$\xi_{\mathrm{I}} = Ae^{-i\omega' t}\,,$$

$$\xi_{\mathrm{II}} = Be^{-i\omega'' t}\,,$$

then upon substitution,

$$-\omega'^2 + \omega_0^2 + \frac{eH}{mc}\,\omega' = 0\,,$$

$$-\omega''^2 + \omega_0^2 - \frac{eH}{mc}\,\omega'' = 0\,.$$

This last system of equations is rigorously solvable, but it is sufficient to expand the roots ω' and ω'':

$$\left.\begin{aligned} \omega' &= \omega_0 + \frac{eH}{2mc} \\ \omega'' &= \omega_0 - \frac{eH}{2mc} \end{aligned}\right\}. \qquad [20.2]$$

The quantity $eH/2mc$ is called the *Larmor frequency.*

Thus, upon turning on the magnetic field, the oscillation of frequency ω_0 has been split into

1. an oscillation parallel to $\boldsymbol{H}$ (the so-called π oscillation), with unchanged frequency ω_0;
2. two oscillations, one right-hand and one left-hand circularly polarized, both perpendicular to $\boldsymbol{H}$ (the so-called σ oscillations), whose frequencies are symmetrically displaced from ω_0 by an amount $eH/2mc$. The oscillation with the higher frequency (for e positive) is left-hand circularly polarized [A-2].

For the case of electrons emitting light, this splitting of the oscillations gives rise to a splitting of the spectral lines. If observations are made transverse to the magnetic field, then the π oscillation appears linearly polarized parallel to $\boldsymbol{H}$ and the σ oscillations appear linearly polarized perpendicular to $\boldsymbol{H}$. (By the polarization direction, we mean here the direction of the *electric* vector.) If observations are made along the direction of the magnetic field, then the π oscillations are not seen at all, while the σ oscillations appear circularly polarized with the higher-frequency oscillation being left-hand circularly polarized. From this, however, it follows that the radiating charge must have a negative sign because, otherwise, if the observation direction were reversed, the higher-frequency oscillation would be associated with the left-hand circular polarization [A-2].

The Zeeman effect permits the determination of e/m (including the sign); the value thus obtained is that for electrons.

The simple Zeeman splitting considered above is not the most general case. For many spectral lines more complicated splittings with several π and σ oscillations appear; this is the so-called anomalous Zeeman effect. This effect is due to the spin of the electron. However, the corresponding theory can be treated only with wave mechanics.

b. Faraday effect

This effect consists of the rotation of the plane of oscillation of light in a magnetic field $\boldsymbol{H}$. The refractive index is different for right-hand and left-hand circularly polarized light.

1. *Light rays parallel to* $\boldsymbol{H}$. The fundamental equation becomes

$$\ddot{\boldsymbol{x}} + \omega_0^2 \boldsymbol{x} = \frac{e}{mc} \dot{\boldsymbol{x}} \times \boldsymbol{H} + \frac{e}{m} \boldsymbol{E}. \tag{20.3}$$

Now,

$$\boldsymbol{P} = Ne\boldsymbol{x},$$

where N is the number of particles per unit volume. We choose $\boldsymbol{H}$ parallel to the z axis. Let the incident wave have angular frequency ω, so that $\boldsymbol{x}$, which will execute forced oscillations, also has this frequency. We make the following substitutions:

$$E_x \pm iE_y = E_{\mathrm{I,II}},$$

$$P_x \pm iP_y = P_{\mathrm{I,II}},$$

from which follows

$$\left(-\omega^2 + \omega_0^2 \pm \frac{eH}{mc}\omega\right) P_{\mathrm{I,II}} = N \frac{e^2}{m} E_{\mathrm{I,II}}.$$

Upon neglecting the corrections between $\boldsymbol{E}'$ and $\boldsymbol{E}$ we have Eq. [16.7]:

$$4\pi \boldsymbol{P} = (n^2 - 1)\boldsymbol{E}.$$

Thus,

$$n_{\mathrm{I}}^2-1=\frac{4\pi(e^2/m)N}{\omega_0^2-\omega^2+(eH/mc)\omega},$$

$$n_{\mathrm{II}}^2-1=\frac{4\pi(e^2/m)N}{\omega_0^2-\omega^2-(eH/mc)\omega},$$

so that

$$n_{\mathrm{I}}^2-n_{\mathrm{II}}^2=4\pi\frac{e^2}{m}N\left\{\frac{1}{\omega_0^2-\omega^2+(eH/mc)\omega}-\frac{1}{\omega_0^2-\omega^2-(eH/mc)\omega}\right\}.$$

With the assumption that ω does not lie close to the eigenfrequency ω_0, this can be simplified (mean value theorem):

$$\left.\begin{aligned} n_{\mathrm{I}}^2-n_{\mathrm{II}}^2&=-4\pi\frac{e^2}{m}N\frac{1}{(\omega_0^2-\omega^2)^2}2\frac{eH}{mc}\omega\\ n_{\mathrm{II}}-n_{\mathrm{I}}&=4\pi\frac{e^2}{m}N\frac{1}{(\omega_0^2-\omega^2)^2}\frac{\omega}{n}\frac{eH}{mc}\end{aligned}\right\}. \qquad [20.4]$$

Now, however, n_{I} and n_{II} are by hypothesis just the refractive indices for the two circular components of the linearly polarized incident wave. Thus, in accordance with Eq. [19.1], the angle of rotation χ of the plane of oscillation becomes

$$\chi=vlH,$$

with

$$v=\frac{4\pi(e^2/m)N}{(\omega_0^2-\omega^2)^2}\frac{e}{mc^2}\frac{\omega^2}{2n}. \qquad [20.5]$$

The quantity v is called Verdet's constant. For the case of several particles in a molecule one must, of course, use the following sum for v:

$$v=\sum_k\frac{4\pi(e_k^2/m_k)N_k}{(\omega_k^2-\omega^2)^2}\frac{e_k\,\omega^2}{m_kc^2}\frac{1}{2n}. \qquad [20.6]$$

2. *Transverse incidence.* In this case birefringence will be observed: one oscillation parallel to $\boldsymbol{H}$ and one perpendicular to $\boldsymbol{H}$.

From the technical point of view, the Faraday effect is much easier to detect than the Zeeman effect, although from the theoretical standpoint the Zeeman effect is simpler. The Zeeman effect consists merely of free oscillations of electrons which are influenced by a static magnetic field, while for the Faraday effect forced oscillations under otherwise identical conditions occur.

Supplementary Bibliography

M. BORN, *Optik* (Springer Verlag, Berlin, Heidelberg, New York 1965).

M. BORN and E. WOLF, *Principles of Optics* (Pergamon Press New York, 1965).

R. DITCHBURN, *Light* (Interscience, New York, 1963).

A. C. HARDY and F. H. PERRIN, *The Principles of Optics* (McGraw Hill, New York, 1932).

F. A. JENKINS and H. E. WHITE, *Fundamentals of Optics* (McGraw Hill, New York, 1957).

B. B. ROSSI, *Optics* (Addison-Wesley, Reading, Massachusetts 1957).

Appendix. Comments by the Editor

[A-1] (p. 23). This is a Legendre transformation from $\boldsymbol{x}$, $\boldsymbol{x}^0$ to $\boldsymbol{p}$, $\boldsymbol{p}^0$ with $\mathrm{d}S = \boldsymbol{p}\cdot\mathrm{d}\boldsymbol{x} - \boldsymbol{p}^0\cdot\mathrm{d}\boldsymbol{x}^0$. The latter form is a special case of Eq. [3.12].

[A-2] (pp. 31, 68, 145). The case $a = 0$ gives a clockwise rotation which by convention is called right-hand circular polarization; $b = 0$ gives a counterclockwise rotation called left-hand circular polarization. Convention fixes the sense of rotation as seen when looking against the direction of light propagation. It also fixes the plane of polarization to be spanned by the magnetic field vector and the wave vector. See M. Born, *Optik*, p. 24 (Supplementary Bibliography).

[A-3] (p. 34). With the advent of high-power lasers (coherent light sources) nonlinear optics has become a fascinating reality (excitation of harmonics, light-light scattering, etc.). See, for example, N. Bloembergen, *Nonlinear Optics* (Benjamin, New York, 1965).

[A-4] (p. 62). It follows from Eqs. [7.8], [7.8′] with $\alpha_1 = \beta_0 = \beta_1 = 0$, $\alpha_0 = \sin\varphi$ that

$$u = \text{constant} \sum_{p=0}^{m-1} \frac{e^{ik(R_{0p}+R_{1p})}}{R_{0p}R_{1p}} \int_{(p-1)a}^{pa} e^{-ik\xi\sin\varphi}\,\mathrm{d}\xi\ ,$$

where a is the height of the steps and $R_{1p} = R_1 + npd$, $R_{0p} = R_0 - pd\cos\varphi$, so that, for small φ, $R_{1p} + R_{0p} \simeq R_1$

$+ R_0 + pd(n-1)$. (The denominator may be approximated by R_0R_1.) Integration and summation yield

$$u = \text{constant}\,\frac{1-e^{ika\sin\varphi}}{\sin\varphi}\,\frac{1-e^{ikm[d(n-1)-a\sin\varphi]}}{1-e^{ik[d(n-1)-a\sin\varphi]}}\,.$$

The first factor vanishes for $(a/\lambda)\sin\varphi = \pm 1$ which for $a \gg \lambda = 2\pi/k$ limits the refraction angle φ to small values. The principal maxima are given by the simultaneous vanishing of numerator and denominator of the second factor,

$$m[d(n-1) - a\sin\varphi] = N_0\lambda\,, \qquad N_0 \text{ integer};$$

$$d(n-1) - a\sin\varphi = N\lambda\,, \qquad N \text{ integer}.$$

Within the interval $|(a/\lambda)\sin\varphi| < 1$ there is only one principal maximum $N \cong (d/\lambda)(n-1)$. The zero next to it is displaced by $\Delta\varphi$ as given by

$$-a\,\Delta\varphi = \left(\frac{N_0 \pm 1}{m} - N\right)\lambda = \frac{\lambda}{m}\,.$$

On the other hand a change of wavelength corresponds to a displacement $\delta\varphi$ given by

$$d\,\frac{\mathrm{d}n}{\mathrm{d}\lambda}\,\delta\lambda - a\,\delta\varphi = N\delta\lambda \cong \frac{d}{\lambda}\,(n-1)\,\delta\lambda\,.$$

The resolving power is determined by the condition $\Delta\varphi = \delta\varphi$ and is given by

$$A \equiv \frac{\lambda}{\delta\lambda} = md\left(\frac{n-1}{\lambda} - \frac{\mathrm{d}n}{\mathrm{d}\lambda}\right).$$

[A-5] (p. 108). The situation is as presented in figure A.1. See M. Born, *Optik*, p. 238 (Supplementary Bibliography).

[A-6) (p. 113). This is the f-sum rule of Thomas and Kuhn (see M. Born, *Optik*, pp. 468, 525, Supplementary Bibliography). In quantum mechanics the f-sum rule is a consequence of the commutation relation between the

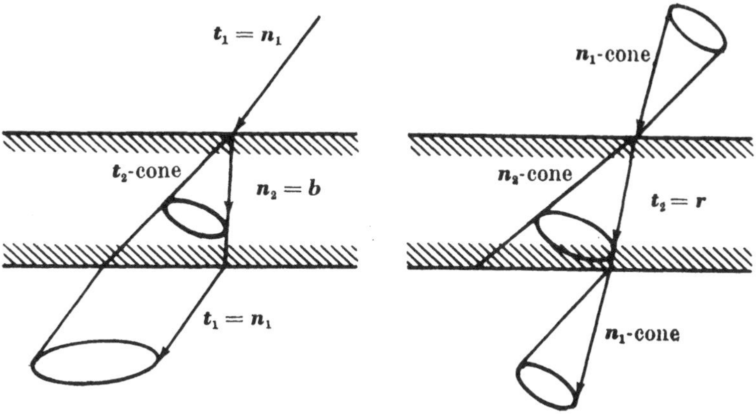

Figure A.1

coordinates x_k and the velocity components $\dot{x}_l$, $x_k\dot{x}_l - \dot{x}_l x_k = i\omega_0 \delta_{kl}$.

[A-7] (p. 138). W. Kuhn, *Z. physik. Chemie* **B4**, 14 (1929). See M. Born, *Optik*, p. 527 (Supplementary Bibliography).

Index

Absorption, 25, 117
Amplitude, complex, 29
Aperture, 48, 49, 53, 55, 60, 62
numerical, 61
Axes,
optical (optic), 90, 101
principal, 85
ray, 105

Babinet's theorem, 49
Bessel function, 57
Bessel's inequality, 32
Binormals, 101
Biradials, 105
Birefringence, 108, 147
Boundary conditions, 46, 47, 48, 50
Brewster's law, 75
Brun's law, 11

Canonical equations, 15, 18, 20
inhomogeneous form of, 22
Charged particles, 109
Circular frequency, 28
Clausius-Mosotti equation, 112
Coherent light sources, 61, 151
Coherent wave trains, 43
Completeness relation, 32
Conductivity, electrical, 77, 80
Cone,
ray, 104
wave, 105
Conservation of energy, 16
for electromagnetic waves, 69, 85, 114

Cornu spiral, 64
Crystal optics, 16, 84–108
Crystals, 20
optically active, 137
uniaxial, 90
Current density, 109

Damped oscillations, 118
Damping, radiation, 121
Damping constant, 79
Damping force, 117
Diaphragm, 49, 51, 52, 55. *See also* Aperture
Dielectric constant,
complex, 78–82
principal, 86
Dielectric constant tensor, 84–86
Differential equations for light rays, 10, 19, 22
Diffraction,
from circular aperture, 55–57
Fraunhofer, 53
Fresnel, 53, 62–66
from grating, 57–59
by slit, 54–55
Diffraction angle, 48, 51, 55, 57, 59
Diffraction pattern, 58, 59, 61
Dipole moment,
electric, 109
electric, of optically active crystals, 140
magnetic, of optically active crystals, 142
Discontinuity of *E* at surface, 70, 71

Dispersion law, 34, 35, 38, 117
for x rays, 128
Doppler effect, 133
Dual equations, 95, 96
Dual variables, 92

Eikonal, 8
angular, 25
Eikonal equation, 8
Electrons, 113, 119, 127–131, 145
Electron, spin of, 146
Emission, 25
Energy, scattered, 125
Energy conservation, 16
for electromagnetic waves, 69, 85, 114
Energy density,
of electromagnetic field, 85
of material oscillators, 115
total, 116
Energy transport, direction of, 93
Euler, L., 28
Euler's homogenity relation, 16, 99
Extremal curve, 5
Extremal principle in canonical form, 15
Extremal property, 2

Faraday effect, 146–148
Fermat's principle, 3, 4, 6, 7, 14
Fluctuations,
density, 124
theory of, 126
Focal plane, 61
Fourier inversion formula, 35
Fourier's theorem, 31, 35
Fraunhofer diffraction, 53, 54–62
Fresnel, A. J., 28
Fresnel diffraction, 53, 62–66
Fresnel formulas, 6
Fresnel integral, 63
Fresnel mirror-experiment, 43
Fresnel's formulae,
for absorbing media, 79–83
for nonabsorbing media, 73–77
Frequency,
circular, 28
Larmor, 145
f-sum rule, 152–153

Grating, 57, 58, 60, 61
step-, 62
Green's theorem, 46
Group velocity, 37, 98, 115, 116

Hamilton, W. R., 10
Hamilton's theory,
homogeneous form of, 14
inhomogeneous form of, 21
Harmonic oscillation, 28
Hertz oscillator, 123
energy radiation of, 124
Hertz vector, 122
Huygens, C., 106
Huygens's principle, 9, 20

Illuminance, 26, 42
Image plane, 61
Image space, 11
Index of refraction (refractive index), 2, 40, 111
of crystals (double root), 88, 100, 101
in Faraday effect, 146
of metals (complex), 77–83, 120
in optical activity, 133, 136
variable, 4, 28
Intensity, 29, 43, 57, 58, 74, 125
incident, 75, 130, 131
of illumination, 26
reflected, 75, 80
refracted, 75
of scattered x rays, 130
specific, of light, 26
time-averaged, 30
Intensity maxima, 44
Intensity ratio, 65
Interference, 28, 42, 44, 45, 67

Kepler, law of, 25. *See also* Lambert, law of
Kirchhoff-Clausius law, 12

Kirchhoff, solution of, 45–53
Kuhn, *f*-sum rule of Thomas and, 152–153
Kuhn's model, 138–143

Lagrangian multipliers, method of, 3, 15, 90
Lambert, law of, 25, 26
Larmor frequency, 145
Lasers, 151
Legendre transformation, 151
Lenses, 13, 53
Light,
  circularly polarized, 133, 134, 146
  corpuscular theory of, 28
  elliptically polarized, 77, 81
  linearly polarized, 75, 76, 77, 80
  polarized, 67
  unpolarized, 67, 75, 131
Light intensity, specific, 26
Light vector, 67, 68
Lorentz force, 144
Lorentz's theory of scattering, 124

Magnetic field (static), 143, 146
Malus's law, 9
Mass absorption coefficient, 128
Maxwell equations, 68, 110, 134
  in crystal optics, 84
  for periodic processes, 71, 77, 86, 121
Mechanical momentum, 16
Mechanical problem, associated, 14
Mechanical waves, 34
Media,
  absorbing (metals), 77–83, 119–121
  anisotropic, 141
  dispersion-free, 39
  homogeneous, 1, 124
  inhomogeneous, 41
  isotropic, 1, 34, 39, 41, 42, 45, 71, 116, 138, 141
  nonabsorbing, 71–77
Metals. *See* Media, absorbing (metals)
Michelson's transmission echelon. *See* Grating, step-
Microscope, 59
Mirror,
  Fresnel, 43
  perfect, 82
Molecules, not mirror-symmetric, 133

Newton's rings, 45
Nodes of E and H in standing waves, 83
Nonlinear optics, 151

Object,
  illuminated, 60
  luminous, 59
Object space, 11
Optical activity, 133
Optical axes, 90, 101
Optical constant for metals, 82, 120
Optical instruments, 62
Optically active crystals, 137
Optically active molecule, 138
Optically uniaxial crystals, 90
Optical path, 3, 4, 6, 107
Oscillation,
  damped, 118
  harmonic, 28
  π (parallel to H), 145
  σ (perpendicular to H), 145
  vectorial, 30
Oscillator,
  Hertz, 123
  quasi-elastic, 110
Oscillator strength, 113

Parseval's formula, 33, 36. *See also* Completeness relation
Pedal surface, 97
Phase,
  continuity of, 40, 72
  propagation of, 93
Phase velocity, 34, 40, 68, 71, 86, 88, 110
Photometric brightness. *See* Light intensity, specific
Plane,
  of incidence, 2, 72, 73, 79, 80, 81, 82

Plane (*continued*)
of polarization, 68, 76, 81, 83, 151
Polarization, 67, 77, 79
circular, 133, 145, 151
electric, 111, 124
plane of, 68, 76
for plane waves, 69, 79
Potentials,
scalar, 121
vector, 121
Poynting vector, 74, 77, 84, 91, 93, 114
Principal angle of incidence, 81
Principal axes, 85
Principal dielectric constants, 86
Principal maxima, 58, 59, 61, 152
Principal section, 88, 89

Quantum-mechanical model, 120, 131
Quantum (wave) mechanics, 146, 153

Radiation, energy, 124
Radiation condition, 47
Radiation damping, 121
Radiator, isotropic, 26
Ray,
extraordinary, 89
ordinary, 88, 89
parallel, 53
reflected, 2, 75
refracted, 2, 75
Ray axes, 105
Ray cone, 104
Ray direction, 20, 93, 97–107
Ray equation, 10, 16, 19, 22
Ray index, 91, 97, 98
Ray surface, 98
Ray variables, 91
Ray vector, 91
Ray velocity, 91
Reflection,
law of, 2, 4, 41
total, 75, 76, 77
Refraction,
double, 108
index of (refractive index), 2, 4, 28, 40, 79, 86, 111, 120, 133, 136, 146
law of, 2, 4, 41
Refractivity, molar, 113
Resolving power, 59–62, 152
Rotation of plane of oscillation,
angle of, 133, 147
in Faraday effect, 146
in optical activity, 133, 136

Scattered energy, 125
Scattering, light-light, 151
Scattering angle, 132
Scattering coefficient, 126, 127
Screens, 46, 48, 49
Shadow boundary, 51, 65
Sine condition, 12, 60
Singular directions, 99, 105
Slit, 54, 63
Snell's law, 1, 7
Spectral lines, splitting of, 145, 146
Statistical average, 132
Statistical mechanics, 124, 126
Structure factor, 130
in quantum theory, 132
Superposition principle, linear, 29, 34, 35, 38
Surface,
of constant phase, 86, 107
pedal, 97, 98
ray, 98
wave, 20, 97
Surface wave, 77

Tensor,
permutation, 137
skew-symmetric, 134, 137
Thomas and Kuhn, *f*-sum rule of, 153
Thomson, J. J., 128
Total reflection, 75, 76, 77
limiting angle of, 76
Transversality condition, 18
Transversality of light waves, 28, 67

Variational principle, 11, 16, 17, 21

Vectors,
axial, 134
polar, 134
Verdet's constant, 147

Wave equation, 38
general solution of, 39
Kirchhoff's solution of, 48, 49
time-independent, 71
Wave,
damped, 79
electromagnetic, 69, 74
harmonic, 43, 45
incident, 71, 72, 73, 74
mechanical, 34
outgoing, 47
plane, 33, 39, 69, 71, 72, 80, 86, 107, 135
reflected, 72, 73, 74, 76, 82
refracted, 73, 74, 76, 79, 80
scalar, 67
spherical, 39
standing, 82
surface, 77
transverse vector, 67
Wave concept, 28
Wave cone, 105
Wave kinematics, 28, 38
Wavelength, 42, 45, 59, 124, 152
Wave normal, 20, 34
Wave packet, 35, 37, 45, 93
Wave surface, 20, 97

X rays, 111, 124, 128
scattering of, 127–133

Young, T., 28

Zeeman effect, 143–146

A CATALOG OF SELECTED

DOVER BOOKS

IN SCIENCE AND MATHEMATICS

Astronomy

BURNHAM'S CELESTIAL HANDBOOK, Robert Burnham, Jr. Thorough guide to the stars beyond our solar system. Exhaustive treatment. Alphabetical by constellation: Andromeda to Cetus in Vol. 1; Chamaeleon to Orion in Vol. 2; and Pavo to Vulpecula in Vol. 3. Hundreds of illustrations. Index in Vol. 3. 2,000pp. 6⅛ x 9¼.
Vol. I: 23567-X
Vol. II: 23568-8
Vol. III: 23673-0

EXPLORING THE MOON THROUGH BINOCULARS AND SMALL TELESCOPES, Ernest H. Cherrington, Jr. Informative, profusely illustrated guide to locating and identifying craters, rills, seas, mountains, other lunar features. Newly revised and updated with special section of new photos. Over 100 photos and diagrams. 240pp. 8¼ x 11. 24491-1

THE EXTRATERRESTRIAL LIFE DEBATE, 1750–1900, Michael J. Crowe. First detailed, scholarly study in English of the many ideas that developed from 1750 to 1900 regarding the existence of intelligent extraterrestrial life. Examines ideas of Kant, Herschel, Voltaire, Percival Lowell, many other scientists and thinkers. 16 illustrations. 704pp. 5⅜ x 8½. 40675-X

THEORIES OF THE WORLD FROM ANTIQUITY TO THE COPERNICAN REVOLUTION, Michael J. Crowe. Newly revised edition of an accessible, enlightening book recreates the change from an earth-centered to a sun-centered conception of the solar system. 242pp. 5⅜ x 8½. 41444-2

A HISTORY OF ASTRONOMY, A. Pannekoek. Well-balanced, carefully reasoned study covers such topics as Ptolemaic theory, work of Copernicus, Kepler, Newton, Eddington's work on stars, much more. Illustrated. References. 521pp. 5⅜ x 8½. 65994-1

A COMPLETE MANUAL OF AMATEUR ASTRONOMY: Tools and Techniques for Astronomical Observations, P. Clay Sherrod with Thomas L. Koed. Concise, highly readable book discusses: selecting, setting up and maintaining a telescope; amateur studies of the sun; lunar topography and occultations; observations of Mars, Jupiter, Saturn, the minor planets and the stars; an introduction to photoelectric photometry; more. 1981 ed. 124 figures. 26 halftones. 37 tables. 335pp. 6½ x 9¼. 42820-6

AMATEUR ASTRONOMER'S HANDBOOK, J. B. Sidgwick. Timeless, comprehensive coverage of telescopes, mirrors, lenses, mountings, telescope drives, micrometers, spectroscopes, more. 189 illustrations. 576pp. 5⅝ x 8¼. (Available in U.S. only.) 24034-7

STARS AND RELATIVITY, Ya. B. Zel'dovich and I. D. Novikov. Vol. 1 of *Relativistic Astrophysics* by famed Russian scientists. General relativity, properties of matter under astrophysical conditions, stars, and stellar systems. Deep physical insights, clear presentation. 1971 edition. References. 544pp. 5⅝ x 8¼. 69424-0

Chemistry

THE SCEPTICAL CHYMIST: The Classic 1661 Text, Robert Boyle. Boyle defines the term "element," asserting that all natural phenomena can be explained by the motion and organization of primary particles. 1911 ed. viii+232pp. 5⅜ x 8½. 42825-7

RADIOACTIVE SUBSTANCES, Marie Curie. Here is the celebrated scientist's doctoral thesis, the prelude to her receipt of the 1903 Nobel Prize. Curie discusses establishing atomic character of radioactivity found in compounds of uranium and thorium; extraction from pitchblende of polonium and radium; isolation of pure radium chloride; determination of atomic weight of radium; plus electric, photographic, luminous, heat, color effects of radioactivity. ii+94pp. 5⅜ x 8½. 42550-9

CHEMICAL MAGIC, Leonard A. Ford. Second Edition, Revised by E. Winston Grundmeier. Over 100 unusual stunts demonstrating cold fire, dust explosions, much more. Text explains scientific principles and stresses safety precautions. 128pp. 5⅜ x 8½. 67628-5

THE DEVELOPMENT OF MODERN CHEMISTRY, Aaron J. Ihde. Authoritative history of chemistry from ancient Greek theory to 20th-century innovation. Covers major chemists and their discoveries. 209 illustrations. 14 tables. Bibliographies. Indices. Appendices. 851pp. 5⅜ x 8½. 64235-6

CATALYSIS IN CHEMISTRY AND ENZYMOLOGY, William P. Jencks. Exceptionally clear coverage of mechanisms for catalysis, forces in aqueous solution, carbonyl- and acyl-group reactions, practical kinetics, more. 864pp. 5⅜ x 8½. 65460-5

ELEMENTS OF CHEMISTRY, Antoine Lavoisier. Monumental classic by founder of modern chemistry in remarkable reprint of rare 1790 Kerr translation. A must for every student of chemistry or the history of science. 539pp. 5⅜ x 8½. 64624-6

THE HISTORICAL BACKGROUND OF CHEMISTRY, Henry M. Leicester. Evolution of ideas, not individual biography. Concentrates on formulation of a coherent set of chemical laws. 260pp. 5⅜ x 8½. 61053-5

A SHORT HISTORY OF CHEMISTRY, J. R. Partington. Classic exposition explores origins of chemistry, alchemy, early medical chemistry, nature of atmosphere, theory of valency, laws and structure of atomic theory, much more. 428pp. 5⅜ x 8½. (Available in U.S. only.) 65977-1

GENERAL CHEMISTRY, Linus Pauling. Revised 3rd edition of classic first-year text by Nobel laureate. Atomic and molecular structure, quantum mechanics, statistical mechanics, thermodynamics correlated with descriptive chemistry. Problems. 992pp. 5⅜ x 8½. 65622-5

FROM ALCHEMY TO CHEMISTRY, John Read. Broad, humanistic treatment focuses on great figures of chemistry and ideas that revolutionized the science. 50 illustrations. 240pp. 5⅜ x 8½. 28690-8

Engineering

DE RE METALLICA, Georgius Agricola. The famous Hoover translation of greatest treatise on technological chemistry, engineering, geology, mining of early modern times (1556). All 289 original woodcuts. 638pp. 6¾ x 11. 60006-8

FUNDAMENTALS OF ASTRODYNAMICS, Roger Bate et al. Modern approach developed by U.S. Air Force Academy. Designed as a first course. Problems, exercises. Numerous illustrations. 455pp. 5⅜ x 8½. 60061-0

DYNAMICS OF FLUIDS IN POROUS MEDIA, Jacob Bear. For advanced students of ground water hydrology, soil mechanics and physics, drainage and irrigation engineering, and more. 335 illustrations. Exercises, with answers. 784pp. 6⅛ x 9¼. 65675-6

THEORY OF VISCOELASTICITY (Second Edition), Richard M. Christensen. Complete, consistent description of the linear theory of the viscoelastic behavior of materials. Problem-solving techniques discussed. 1982 edition. 29 figures. xiv+364pp. 6⅛ x 9¼. 42880-X

MECHANICS, J. P. Den Hartog. A classic introductory text or refresher. Hundreds of applications and design problems illuminate fundamentals of trusses, loaded beams and cables, etc. 334 answered problems. 462pp. 5⅜ x 8½. 60754-2

MECHANICAL VIBRATIONS, J. P. Den Hartog. Classic textbook offers lucid explanations and illustrative models, applying theories of vibrations to a variety of practical industrial engineering problems. Numerous figures. 233 problems, solutions. Appendix. Index. Preface. 436pp. 5⅜ x 8½. 64785-4

STRENGTH OF MATERIALS, J. P. Den Hartog. Full, clear treatment of basic material (tension, torsion, bending, etc.) plus advanced material on engineering methods, applications. 350 answered problems. 323pp. 5⅜ x 8½. 60755-0

A HISTORY OF MECHANICS, René Dugas. Monumental study of mechanical principles from antiquity to quantum mechanics. Contributions of ancient Greeks, Galileo, Leonardo, Kepler, Lagrange, many others. 671pp. 5⅜ x 8½. 65632-2

STABILITY THEORY AND ITS APPLICATIONS TO STRUCTURAL MECHANICS, Clive L. Dym. Self-contained text focuses on Koiter postbuckling analyses, with mathematical notions of stability of motion. Basing minimum energy principles for static stability upon dynamic concepts of stability of motion, it develops asymptotic buckling and postbuckling analyses from potential energy considerations, with applications to columns, plates, and arches. 1974 ed. 208pp. 5⅜ x 8½. 42541-X

METAL FATIGUE, N. E. Frost, K. J. Marsh, and L. P. Pook. Definitive, clearly written, and well-illustrated volume addresses all aspects of the subject, from the historical development of understanding metal fatigue to vital concepts of the cyclic stress that causes a crack to grow. Includes 7 appendixes. 544pp. 5⅜ x 8½. 40927-9

Mathematics

FUNCTIONAL ANALYSIS (Second Corrected Edition), George Bachman and Lawrence Narici. Excellent treatment of subject geared toward students with background in linear algebra, advanced calculus, physics, and engineering. Text covers introduction to inner-product spaces, normed, metric spaces, and topological spaces; complete orthonormal sets, the Hahn-Banach Theorem and its consequences, and many other related subjects. 1966 ed. 544pp. 6⅛ x 9¼. 40251-7

ASYMPTOTIC EXPANSIONS OF INTEGRALS, Norman Bleistein & Richard A. Handelsman. Best introduction to important field with applications in a variety of scientific disciplines. New preface. Problems. Diagrams. Tables. Bibliography. Index. 448pp. 5⅜ x 8½. 65082-0

VECTOR AND TENSOR ANALYSIS WITH APPLICATIONS, A. I. Borisenko and I. E. Tarapov. Concise introduction. Worked-out problems, solutions, exercises. 257pp. 5⅝ x 8¼. 63833-2

THE ABSOLUTE DIFFERENTIAL CALCULUS (CALCULUS OF TENSORS), Tullio Levi-Civita. Great 20th-century mathematician's classic work on material necessary for mathematical grasp of theory of relativity. 452pp. 5⅝ x 8¼. 63401-9

AN INTRODUCTION TO ORDINARY DIFFERENTIAL EQUATIONS, Earl A. Coddington. A thorough and systematic first course in elementary differential equations for undergraduates in mathematics and science, with many exercises and problems (with answers). Index. 304pp. 5⅜ x 8½. 65942-9

FOURIER SERIES AND ORTHOGONAL FUNCTIONS, Harry F. Davis. An incisive text combining theory and practical example to introduce Fourier series, orthogonal functions and applications of the Fourier method to boundary-value problems. 570 exercises. Answers and notes. 416pp. 5⅜ x 8½. 65973-9

COMPUTABILITY AND UNSOLVABILITY, Martin Davis. Classic graduate-level introduction to theory of computability, usually referred to as theory of recurrent functions. New preface and appendix. 288pp. 5⅜ x 8½. 61471-9

ASYMPTOTIC METHODS IN ANALYSIS, N. G. de Bruijn. An inexpensive, comprehensive guide to asymptotic methods–the pioneering work that teaches by explaining worked examples in detail. Index. 224pp. 5⅜ x 8½ 64221-6

APPLIED COMPLEX VARIABLES, John W. Dettman. Step-by-step coverage of fundamentals of analytic function theory–plus lucid exposition of five important applications: Potential Theory; Ordinary Differential Equations; Fourier Transforms; Laplace Transforms; Asymptotic Expansions. 66 figures. Exercises at chapter ends. 512pp. 5⅜ x 8½. 64670-X

INTRODUCTION TO LINEAR ALGEBRA AND DIFFERENTIAL EQUATIONS, John W. Dettman. Excellent text covers complex numbers, determinants, orthonormal bases, Laplace transforms, much more. Exercises with solutions. Undergraduate level. 416pp. 5⅜ x 8½. 65191-6

CALCULUS OF VARIATIONS WITH APPLICATIONS, George M. Ewing. Applications-oriented introduction to variational theory develops insight and promotes understanding of specialized books, research papers. Suitable for advanced undergraduate/graduate students as primary, supplementary text. 352pp. 5⅜ x 8½. 64856-7

COMPLEX VARIABLES, Francis J. Flanigan. Unusual approach, delaying complex algebra till harmonic functions have been analyzed from real variable viewpoint. Includes problems with answers. 364pp. 5⅜ x 8½. 61388-7

AN INTRODUCTION TO THE CALCULUS OF VARIATIONS, Charles Fox. Graduate-level text covers variations of an integral, isoperimetrical problems, least action, special relativity, approximations, more. References. 279pp. 5⅜ x 8½. 65499-0

COUNTEREXAMPLES IN ANALYSIS, Bernard R. Gelbaum and John M. H. Olmsted. These counterexamples deal mostly with the part of analysis known as "real variables." The first half covers the real number system, and the second half encompasses higher dimensions. 1962 edition. xxiv+198pp. 5⅜ x 8½. 42875-3

CATASTROPHE THEORY FOR SCIENTISTS AND ENGINEERS, Robert Gilmore. Advanced-level treatment describes mathematics of theory grounded in the work of Poincaré, R. Thom, other mathematicians. Also important applications to problems in mathematics, physics, chemistry, and engineering. 1981 edition. References. 28 tables. 397 black-and-white illustrations. xvii+666pp. 6⅛ x 9¼. 67539-4

INTRODUCTION TO DIFFERENCE EQUATIONS, Samuel Goldberg. Exceptionally clear exposition of important discipline with applications to sociology, psychology, economics. Many illustrative examples; over 250 problems. 260pp. 5⅜ x 8½. 65084-7

NUMERICAL METHODS FOR SCIENTISTS AND ENGINEERS, Richard Hamming. Classic text stresses frequency approach in coverage of algorithms, polynomial approximation, Fourier approximation, exponential approximation, other topics. Revised and enlarged 2nd edition. 721pp. 5⅜ x 8½. 65241-6

INTRODUCTION TO NUMERICAL ANALYSIS (2nd Edition), F. B. Hildebrand. Classic, fundamental treatment covers computation, approximation, interpolation, numerical differentiation and integration, other topics. 150 new problems. 669pp. 5⅜ x 8½. 65363-3

THREE PEARLS OF NUMBER THEORY, A. Y. Khinchin. Three compelling puzzles require proof of a basic law governing the world of numbers. Challenges concern van der Waerden's theorem, the Landau-Schnirelmann hypothesis and Mann's theorem, and a solution to Waring's problem. Solutions included. 64pp. 5⅜ x 8½. 40026-3

THE PHILOSOPHY OF MATHEMATICS: An Introductory Essay, Stephan Körner. Surveys the views of Plato, Aristotle, Leibniz & Kant concerning propositions and theories of applied and pure mathematics. Introduction. Two appendices. Index. 198pp. 5⅜ x 8½. 25048-2

INTRODUCTORY REAL ANALYSIS, A.N. Kolmogorov, S. V. Fomin. Translated by Richard A. Silverman. Self-contained, evenly paced introduction to real and functional analysis. Some 350 problems. 403pp. 5⅜ x 8½. 61226-0

APPLIED ANALYSIS, Cornelius Lanczos. Classic work on analysis and design of finite processes for approximating solution of analytical problems. Algebraic equations, matrices, harmonic analysis, quadrature methods, more. 559pp. 5⅜ x 8½. 65656-X

AN INTRODUCTION TO ALGEBRAIC STRUCTURES, Joseph Landin. Superb self-contained text covers "abstract algebra": sets and numbers, theory of groups, theory of rings, much more. Numerous well-chosen examples, exercises. 247pp. 5⅜ x 8½. 65940-2

QUALITATIVE THEORY OF DIFFERENTIAL EQUATIONS, V. V. Nemytskii and V.V. Stepanov. Classic graduate-level text by two prominent Soviet mathematicians covers classical differential equations as well as topological dynamics and ergodic theory. Bibliographies. 523pp. 5⅜ x 8½. 65954-2

THEORY OF MATRICES, Sam Perlis. Outstanding text covering rank, nonsingularity and inverses in connection with the development of canonical matrices under the relation of equivalence, and without the intervention of determinants. Includes exercises. 237pp. 5⅜ x 8½. 66810-X

INTRODUCTION TO ANALYSIS, Maxwell Rosenlicht. Unusually clear, accessible coverage of set theory, real number system, metric spaces, continuous functions, Riemann integration, multiple integrals, more. Wide range of problems. Undergraduate level. Bibliography. 254pp. 5⅜ x 8½. 65038-3

MODERN NONLINEAR EQUATIONS, Thomas L. Saaty. Emphasizes practical solution of problems; covers seven types of equations. ". . . a welcome contribution to the existing literature. . . . "–*Math Reviews*. 490pp. 5⅜ x 8½. 64232-1

MATRICES AND LINEAR ALGEBRA, Hans Schneider and George Phillip Barker. Basic textbook covers theory of matrices and its applications to systems of linear equations and related topics such as determinants, eigenvalues, and differential equations. Numerous exercises. 432pp. 5⅜ x 8½. 66014-1

MATHEMATICS APPLIED TO CONTINUUM MECHANICS, Lee A. Segel. Analyzes models of fluid flow and solid deformation. For upper-level math, science, and engineering students. 608pp. 5⅜ x 8½. 65369-2

ELEMENTS OF REAL ANALYSIS, David A. Sprecher. Classic text covers fundamental concepts, real number system, point sets, functions of a real variable, Fourier series, much more. Over 500 exercises. 352pp. 5⅜ x 8½. 65385-4

SET THEORY AND LOGIC, Robert R. Stoll. Lucid introduction to unified theory of mathematical concepts. Set theory and logic seen as tools for conceptual understanding of real number system. 496pp. 5⅜ x 8¼. 63829-4

Math–Decision Theory, Statistics, Probability

ELEMENTARY DECISION THEORY, Herman Chernoff and Lincoln E. Moses. Clear introduction to statistics and statistical theory covers data processing, probability and random variables, testing hypotheses, much more. Exercises. 364pp. 5⅜ x 8½. 65218-1

STATISTICS MANUAL, Edwin L. Crow et al. Comprehensive, practical collection of classical and modern methods prepared by U.S. Naval Ordnance Test Station. Stress on use. Basics of statistics assumed. 288pp. 5⅜ x 8½. 60599-X

SOME THEORY OF SAMPLING, William Edwards Deming. Analysis of the problems, theory, and design of sampling techniques for social scientists, industrial managers, and others who find statistics important at work. 61 tables. 90 figures. xvii +602pp. 5⅜ x 8½. 64684-X

LINEAR PROGRAMMING AND ECONOMIC ANALYSIS, Robert Dorfman, Paul A. Samuelson and Robert M. Solow. First comprehensive treatment of linear programming in standard economic analysis. Game theory, modern welfare economics, Leontief input-output, more. 525pp. 5⅜ x 8½. 65491-5

PROBABILITY: An Introduction, Samuel Goldberg. Excellent basic text covers set theory, probability theory for finite sample spaces, binomial theorem, much more. 360 problems. Bibliographies. 322pp. 5⅜ x 8½. 65252-1

GAMES AND DECISIONS: Introduction and Critical Survey, R. Duncan Luce and Howard Raiffa. Superb nontechnical introduction to game theory, primarily applied to social sciences. Utility theory, zero-sum games, n-person games, decision-making, much more. Bibliography. 509pp. 5⅜ x 8½. 65943-7

INTRODUCTION TO THE THEORY OF GAMES, J. C. C. McKinsey. This comprehensive overview of the mathematical theory of games illustrates applications to situations involving conflicts of interest, including economic, social, political, and military contexts. Appropriate for advanced undergraduate and graduate courses; advanced calculus a prerequisite. 1952 ed. x+372pp. 5⅜ x 8½. 42811-7

FIFTY CHALLENGING PROBLEMS IN PROBABILITY WITH SOLUTIONS, Frederick Mosteller. Remarkable puzzlers, graded in difficulty, illustrate elementary and advanced aspects of probability. Detailed solutions. 88pp. 5⅜ x 8½. 65355-2

PROBABILITY THEORY: A Concise Course, Y. A. Rozanov. Highly readable, self-contained introduction covers combination of events, dependent events, Bernoulli trials, etc. 148pp. 5⅜ x 8¼. 63544-9

STATISTICAL METHOD FROM THE VIEWPOINT OF QUALITY CONTROL, Walter A. Shewhart. Important text explains regulation of variables, uses of statistical control to achieve quality control in industry, agriculture, other areas. 192pp. 5⅜ x 8½. 65232-7

Math–Geometry and Topology

ELEMENTARY CONCEPTS OF TOPOLOGY, Paul Alexandroff. Elegant, intuitive approach to topology from set-theoretic topology to Betti groups; how concepts of topology are useful in math and physics. 25 figures. 57pp. 5⅜ x 8½. 60747-X

COMBINATORIAL TOPOLOGY, P. S. Alexandrov. Clearly written, well-organized, three-part text begins by dealing with certain classic problems without using the formal techniques of homology theory and advances to the central concept, the Betti groups. Numerous detailed examples. 654pp. 5⅜ x 8½. 40179-0

EXPERIMENTS IN TOPOLOGY, Stephen Barr. Classic, lively explanation of one of the byways of mathematics. Klein bottles, Moebius strips, projective planes, map coloring, problem of the Koenigsberg bridges, much more, described with clarity and wit. 43 figures. 210pp. 5⅜ x 8½. 25933-1

CONFORMAL MAPPING ON RIEMANN SURFACES, Harvey Cohn. Lucid, insightful book presents ideal coverage of subject. 334 exercises make book perfect for self-study. 55 figures. 352pp. 5⅝ x 8¼. 64025-6

THE GEOMETRY OF RENÉ DESCARTES, René Descartes. The great work founded analytical geometry. Original French text, Descartes's own diagrams, together with definitive Smith-Latham translation. 244pp. 5⅜ x 8½. 60068-8

PRACTICAL CONIC SECTIONS: The Geometric Properties of Ellipses, Parabolas and Hyperbolas, J. W. Downs. This text shows how to create ellipses, parabolas, and hyperbolas. It also presents historical background on their ancient origins and describes the reflective properties and roles of curves in design applications. 1993 ed. 98 figures. xii+100pp. 6½ x 9¼. 42876-1

THE THIRTEEN BOOKS OF EUCLID'S ELEMENTS, translated with introduction and commentary by Thomas L. Heath. Definitive edition. Textual and linguistic notes, mathematical analysis. 2,500 years of critical commentary. Unabridged. 1,414pp. 5⅜ x 8½. Three-vol. set. Vol. I: 60088-2 Vol. II: 60089-0 Vol. III: 60090-4

GEOMETRY OF COMPLEX NUMBERS, Hans Schwerdtfeger. Illuminating, widely praised book on analytic geometry of circles, the Moebius transformation, and two-dimensional non-Euclidean geometries. 200pp. 5⅝ x 8¼. 63830-8

DIFFERENTIAL GEOMETRY, Heinrich W. Guggenheimer. Local differential geometry as an application of advanced calculus and linear algebra. Curvature, transformation groups, surfaces, more. Exercises. 62 figures. 378pp. 5⅜ x 8½. 63433-7

CURVATURE AND HOMOLOGY: Enlarged Edition, Samuel I. Goldberg. Revised edition examines topology of differentiable manifolds; curvature, homology of Riemannian manifolds; compact Lie groups; complex manifolds; curvature, homology of Kaehler manifolds. New Preface. Four new appendixes. 416pp. 5⅜ x 8½. 40207-X

History of Math

THE WORKS OF ARCHIMEDES, Archimedes (T. L. Heath, ed.). Topics include the famous problems of the ratio of the areas of a cylinder and an inscribed sphere; the measurement of a circle; the properties of conoids, spheroids, and spirals; and the quadrature of the parabola. Informative introduction. clxxxvi+326pp; supplement, 52pp. 5⅜ x 8½. 42084-1

A SHORT ACCOUNT OF THE HISTORY OF MATHEMATICS, W. W. Rouse Ball. One of clearest, most authoritative surveys from the Egyptians and Phoenicians through 19th-century figures such as Grassman, Galois, Riemann. Fourth edition. 522pp. 5⅜ x 8½. 20630-0

THE HISTORY OF THE CALCULUS AND ITS CONCEPTUAL DEVELOPMENT, Carl B. Boyer. Origins in antiquity, medieval contributions, work of Newton, Leibniz, rigorous formulation. Treatment is verbal. 346pp. 5⅜ x 8½. 60509-4

THE HISTORICAL ROOTS OF ELEMENTARY MATHEMATICS, Lucas N. H. Bunt, Phillip S. Jones, and Jack D. Bedient. Fundamental underpinnings of modern arithmetic, algebra, geometry, and number systems derived from ancient civilizations. 320pp. 5⅜ x 8½. 25563-8

A HISTORY OF MATHEMATICAL NOTATIONS, Florian Cajori. This classic study notes the first appearance of a mathematical symbol and its origin, the competition it encountered, its spread among writers in different countries, its rise to popularity, its eventual decline or ultimate survival. Original 1929 two-volume edition presented here in one volume. xxviii+820pp. 5⅜ x 8½. 67766-4

GAMES, GODS & GAMBLING: A History of Probability and Statistical Ideas, F. N. David. Episodes from the lives of Galileo, Fermat, Pascal, and others illustrate this fascinating account of the roots of mathematics. Features thought-provoking references to classics, archaeology, biography, poetry. 1962 edition. 304pp. 5⅜ x 8½. (Available in U.S. only.) 40023-9

OF MEN AND NUMBERS: The Story of the Great Mathematicians, Jane Muir. Fascinating accounts of the lives and accomplishments of history's greatest mathematical minds–Pythagoras, Descartes, Euler, Pascal, Cantor, many more. Anecdotal, illuminating. 30 diagrams. Bibliography. 256pp. 5⅜ x 8½. 28973-7

HISTORY OF MATHEMATICS, David E. Smith. Nontechnical survey from ancient Greece and Orient to late 19th century; evolution of arithmetic, geometry, trigonometry, calculating devices, algebra, the calculus. 362 illustrations. 1,355pp. 5⅜ x 8½. Two-vol. set. Vol. I: 20429-4 Vol. II: 20430-8

A CONCISE HISTORY OF MATHEMATICS, Dirk J. Struik. The best brief history of mathematics. Stresses origins and covers every major figure from ancient Near East to 19th century. 41 illustrations. 195pp. 5⅜ x 8½. 60255-9

Physics

OPTICAL RESONANCE AND TWO-LEVEL ATOMS, L. Allen and J. H. Eberly. Clear, comprehensive introduction to basic principles behind all quantum optical resonance phenomena. 53 illustrations. Preface. Index. 256pp. 5⅜ x 8½. 65533-4

QUANTUM THEORY, David Bohm. This advanced undergraduate-level text presents the quantum theory in terms of qualitative and imaginative concepts, followed by specific applications worked out in mathematical detail. Preface. Index. 655pp. 5⅜ x 8½. 65969-0

ATOMIC PHYSICS: 8th edition, Max Born. Nobel laureate's lucid treatment of kinetic theory of gases, elementary particles, nuclear atom, wave-corpuscles, atomic structure and spectral lines, much more. Over 40 appendices, bibliography. 495pp. 5⅜ x 8½. 65984-4

A SOPHISTICATE'S PRIMER OF RELATIVITY, P. W. Bridgman. Geared toward readers already acquainted with special relativity, this book transcends the view of theory as a working tool to answer natural questions: What is a frame of reference? What is a "law of nature"? What is the role of the "observer"? Extensive treatment, written in terms accessible to those without a scientific background. 1983 ed. xlviii+172pp. 5⅜ x 8½. 42549-5

AN INTRODUCTION TO HAMILTONIAN OPTICS, H. A. Buchdahl. Detailed account of the Hamiltonian treatment of aberration theory in geometrical optics. Many classes of optical systems defined in terms of the symmetries they possess. Problems with detailed solutions. 1970 edition. xv+360pp. 5⅜ x 8½. 67597-1

PRIMER OF QUANTUM MECHANICS, Marvin Chester. Introductory text examines the classical quantum bead on a track: its state and representations; operator eigenvalues; harmonic oscillator and bound bead in a symmetric force field; and bead in a spherical shell. Other topics include spin, matrices, and the structure of quantum mechanics; the simplest atom; indistinguishable particles; and stationary-state perturbation theory. 1992 ed. xiv+314pp. 6⅛ x 9¼. 42878-8

LECTURES ON QUANTUM MECHANICS, Paul A. M. Dirac. Four concise, brilliant lectures on mathematical methods in quantum mechanics from Nobel Prize–winning quantum pioneer build on idea of visualizing quantum theory through the use of classical mechanics. 96pp. 5⅜ x 8½. 41713-1

THIRTY YEARS THAT SHOOK PHYSICS: The Story of Quantum Theory, George Gamow. Lucid, accessible introduction to influential theory of energy and matter. Careful explanations of Dirac's anti-particles, Bohr's model of the atom, much more. 12 plates. Numerous drawings. 240pp. 5⅜ x 8½. 24895-X

ELECTRONIC STRUCTURE AND THE PROPERTIES OF SOLIDS: The Physics of the Chemical Bond, Walter A. Harrison. Innovative text offers basic understanding of the electronic structure of covalent and ionic solids, simple metals, transition metals and their compounds. Problems. 1980 edition. 582pp. 6⅛ x 9¼. 66021-4

QUANTUM MECHANICS: Principles and Formalism, Roy McWeeny. Graduate student–oriented volume develops subject as fundamental discipline, opening with review of origins of Schrödinger's equations and vector spaces. Focusing on main principles of quantum mechanics and their immediate consequences, it concludes with final generalizations covering alternative "languages" or representations. 1972 ed. 15 figures. xi+155pp. 5⅜ x 8½. 42829-X

INTRODUCTION TO QUANTUM MECHANICS WITH APPLICATIONS TO CHEMISTRY, Linus Pauling & E. Bright Wilson, Jr. Classic undergraduate text by Nobel Prize winner applies quantum mechanics to chemical and physical problems. Numerous tables and figures enhance the text. Chapter bibliographies. Appendices. Index. 468pp. 5⅜ x 8½. 64871-0

METHODS OF THERMODYNAMICS, Howard Reiss. Outstanding text focuses on physical technique of thermodynamics, typical problem areas of understanding, and significance and use of thermodynamic potential. 1965 edition. 238pp. 5⅜ x 8½. 69445-3

TENSOR ANALYSIS FOR PHYSICISTS, J. A. Schouten. Concise exposition of the mathematical basis of tensor analysis, integrated with well-chosen physical examples of the theory. Exercises. Index. Bibliography. 289pp. 5⅜ x 8½. 65582-2

THE ELECTROMAGNETIC FIELD, Albert Shadowitz. Comprehensive undergraduate text covers basics of electric and magnetic fields, builds up to electromagnetic theory. Also related topics, including relativity. Over 900 problems. 768pp. 5⅝ x 8¼. 65660-8

GREAT EXPERIMENTS IN PHYSICS: Firsthand Accounts from Galileo to Einstein, Morris H. Shamos (ed.). 25 crucial discoveries: Newton's laws of motion, Chadwick's study of the neutron, Hertz on electromagnetic waves, more. Original accounts clearly annotated. 370pp. 5⅜ x 8½. 25346-5

RELATIVITY, THERMODYNAMICS AND COSMOLOGY, Richard C. Tolman. Landmark study extends thermodynamics to special, general relativity; also applications of relativistic mechanics, thermodynamics to cosmological models. 501pp. 5⅜ x 8½. 65383-8

STATISTICAL PHYSICS, Gregory H. Wannier. Classic text combines thermodynamics, statistical mechanics, and kinetic theory in one unified presentation of thermal physics. Problems with solutions. Bibliography. 532pp. 5⅜ x 8½. 65401-X

Paperbound unless otherwise indicated. Available at your book dealer, online at **www.doverpublications.com**, or by writing to Dept. GI, Dover Publications, Inc., 31 East 2nd Street, Mineola, NY 11501. For current price information or for free catalogs (please indicate field of interest), write to Dover Publications or log on to **www.doverpublications.com** and see every Dover book in print. Dover publishes more than 500 books each year on science, elementary and advanced mathematics, biology, music, art, literary history, social sciences, and other areas.